Architects, Sustainability and the Climate Emergency

Living in and with Nature is the single challenge for humanity as climate change rapidly propels us down the path of ecocide and extinction. New visions of ecological sustainability demand that we, literally, build the future in a different way. For this, we need to listen to the voices of intellectual innovation, learn the wisdom of the ancients, and draw energy from those who see, judgc, and act as global calamity unfolds. This fascinating book provides vital insights and critiques at the interfaces of architectural theory and practice, in the process elaborating a contemporary policy ecology of the built form. For anyone concerned about the climate emergency, it is essential reading.

Rob White
Emeritus Distinguished Professor of Criminology, University of Tasmania, Australia

Architects, Sustainability and the Climate Emergency: A Political Ecology

BY

PETER RAISBECK

University of Melbourne, Australia

emerald PUBLISHING

United Kingdom – North America – Japan – India – Malaysia – China

Emerald Publishing Limited
Howard House, Wagon Lane, Bingley BD16 1WA, UK

First edition 2023

British Library Cataloguing in Publication Data
A catalogue record for this book is available from the British Library

ISBN: 978-1-80382-292-1 (Print)
ISBN: 978-1-80382-291-4 (Online)
ISBN: 978-1-80382-293-8 (Epub)

Printed and bound by CPI Group (UK) Ltd, Croydon, CR0 4YY

ISOQAR
REGISTERED
ISOQAR certified Management System, awarded to Emerald for adherence to Environmental standard ISO 14001:2004.

Certificate Number 1985
ISO 14001

INVESTOR IN PEOPLE

Epigraph

They saw the great fire advancing from the Macumba Maka-Wimpa 'fire track' and they got out quickly, traveling to the south-east still looking for more of their own people.

–Luise Hercus. (2009). Murkarra, a landscape nearly forgotten: The Arabana country of the noxious insects, north and northwest of Lake Eyre. In A. L. Hercus & H. J. Koch (Eds.), *Aboriginal placenames: Naming and re-naming the Australian landscape aboriginal history* (p. 257). Monograph 19. ANU E Press and Aboriginal History Incorporated.

Table of Contents

List of Figures

Chapter 3

Chapter 4

Chapter 5

Chapter 6

About the Author

Dr Peter Raisbeck is Associate Professor of Architectural Practice at the Melbourne School of Design University of Melbourne. At MSD he teaches Architectural Practice, Design Activism, and Contemporary Architectural Archives. His research spans the fields of architectural history and the sociology of architectural design and practice. His book, *Architecture as a System: Scavengers, Tribes, Warlords and Megafirms*, (2019) considers architecture as a global system in crisis. His coauthored book with Dr Christine Phillips, *Robin Boyd: Late Works 1960–1971*, (2020) is an extensive survey of the notable Australian architect Robin Boyd's public commissions prior to his death in 1971.

Foreword

When I began writing this book, megafires were burning on the Australian continent I live on, and as I finished writing parts of this book, the land has been inundated with catastrophic floods. In between these events, there has been a global pandemic. These framings of time suggest a finite beginning, middle, and ending to these catastrophic events. However, what has motivated my writing is the intuition that this climate chaos cannot be so easily framed with a chronological ending. All endings suggest we might look beyond to a better time. But I fear that the climate imaginaries conjured by architects, and described in what follows, presume a definitive end to such things – an end to the climate emergency. These imaginaries dictate an enchanted future Anthropocene where human life can go on as usual. Yet, as the climate emergency unfolds, we begin to sense that there are no more neat endings; chaos begets chaos. As described in this book, the magic of architects is an illusion that is even now contributing to future chaos.

Melbourne, March 2022.

Acknowledgments

This book has been written on the lands of the Wurundjeri and Boon Wurrung peoples. I acknowledge the Traditional Owners of this land I have been writing on and pay my respect to their Elders and families past and present.

I would like to thank my wife, Victoria Conners, for her ongoing support through much laughter and the trials of six Melbourne lockdowns. Without this support, this book would not have been possible. I would also like to acknowledge the help of my children, Rachel Raisbeck and Raphael Raisbeck, whose laughter at my humor I will always treasure. Of course, I must also thank those I regard as part of my family, my Christmas day crew: Eric Conners, Carmel Conners, Sarah Conners, and Andrew McIlroy. And of course, Jennifer Klempfner and her partner Rob Larkins, Yolande and George Klempfner. And the amazing Judith Raisbeck as well as my sister Vicky Marney and my brother David Gibson.

Katy Mathers, and her team at Emerald, was also most patient and amazingly encouraging. At Melbourne School of Design, a particular thanks to Dr Kirsten Day for her support in both teaching and research.

At MSD, I would like to thank Julie Willis, Alan Pert, and Paul Walker for their continuing support. Melbourne School of Design is one of the world's great architectural and built environment scholarship centers. This is reflected in the text, where many of my colleagues are great scholars whose ideas have inspired and provoked my thought. Among others: Michele Acuto, Alexei Trundle, Margaret Grose, Robert Crawford, and of course the amazing Helene Frichot – too often I read your work and then forget to chat outside our building's café. My PhD students, Dr Sharon Shafer and Puneeta Thakur.

As the book proceeded, I continued to teach Architectural Practice at Melbourne School of Design, and I am most thankful to Rob Polglase and supporting me in my teaching. I must also thank the beautiful architects who teach in this subject for many conversations: Rebecca Naughtin, Joanna Sych, Redmond Hamlett, Melany Hayes Camilla Tierney, Rowan Brown, Christopher Hewson, Isabelle Legge, Sonia Sarangi, and the indefatigable Dr Norman Day.

I would also like to thank Greg Burgess for entrusting us with their Archive, which in many ways has helped me to think about the final chapters of this book – I hope the next book I write will be about this archive. I would also like to thank Andy Jackson, whose critical comments on my feeble efforts at poetry has helped me prose writing enormously.

I would also like to thank my support team of health professionals, including Dr Katrina Reardon, Dr Sarah Lee, Dr Martyn Delatycki, Dr Ralph Audehm,

Isabelle Aubrey, Rebecca James, Sam Stinson, Amanda Spillare, and Julie Wilson at MND Victoria. Many of whom have enabled me to be highly productive in one way or another.

Gratitude to the other Producers, Noelle Jones and Tanya Spencer. Many friends must also be acknowledged through this journey: Karen Burns and Dean Cass, Tava Darakamaran, Tania Davidge, Michaela Prunotto, Rosanna Verde, Alica Brown and Tamsin Maclean, Christine Canters, Simona Castricum, Kate Tregloan, Amanda Davis and Simon O'Brien, Kirsten Day, John Ting, Peter Hogg, and Helen Slonek. The Hinses' including Karen Hines, Lee Hines and my favorite Hines: Ellyse and Olivia Hines. Susan Diggs and Paul Almond. Rochus Hinkel, Yvette Putra, Christine Phillips, Sue Dance, and Jason Pickford. My discussions with Jason Pickford over the last 20 years concerning climate politics are the bedrock of this book.

As remarked in the acknowledgment to my previous book, often, there are pressures nowadays on academics to avoid the longer and solitary forms of scholarship. Without the assistance of the above people, this contribution to that genre would not have been possible.

Chapter 1

Introduction

In the history of modern architecture, the city has been reimagined by architects in many different guises. This book is how architects have shaped their visions of the city; as their knowledge of the climate emergency has unfolded. At the heart of the book is the question, have architects failed to heed this crisis? As climate fires roar and a pandemic rage, we might ask, do architects sleep too soundly and quietly? Have the climate imaginaries that architects have produced, in response to the science of climate change, served to conceal the climate emergency itself? Modern architecture, as conjured up by its proponents – for all its follies – has always had a focus on the city. Modern architecture was always about and never anything less than an investment in the dream of a future city. There was always a hope that the imagination of modern architects, through their own agency, would shape the cities of the future.

The promises, dreams, and hopes of architects for future cities are now inextricably linked to climate change. Understandably, architects might return to the glories of modern architecture; with its spiritualist promises of social, technological, and, above all, aesthetic freedom. The climate emergency has prompted architects to take up new dreams even as they maintain old rituals and memories. It is the new dreams of architecture in the twenty-first century, constituted by past phantoms and half-remembered images, that are discussed in this book. Each of these dreams incite a euphoria that architecture, with all of its different tributaries of style and theoretical tropes, might yet be something that can offer salvation. In this strange architectural light, between waking and dreaming, what angels, saints, sorcerers, and ghosts have appeared to quell the climate emergency? What monsters, real or imagined, lurk in this landscape of magic and dreams?

The above rhetoric may seem somewhat unorthodox with its hints of deliria and esoteric beings. Nonetheless, magic and superstition are suitable frames when describing how architects have responded and continue to respond to the climate emergency. Architects have proposed so many different cities, and so much magic, and yet, the city that reconciles nature with humanity is still to be built. And so, this book is a detailed examination of architects as a profession and how they have responded and continue to respond to the climate emergency. In doing so, the purpose of this book is to outline a political ecology of architecture.

This political ecology is underscored by the science. The Intergovernmental Panel on Climate Change (IPCC) Special Report 15 on Global Warming was

Architects, Sustainability and the Climate Emergency, 1–28

Published under exclusive licence by Emerald Publishing Limited
doi:10.1108/978-1-80382-291-420221001

published in 2018. This report indicates that human activity has contributed to "approximately 1.0°C of global warming," and the range of this contribution is predicted to be between 0.8°C and 1.2°C (IPCC, 2018, p. 4). If the trend underlying this global warming continues, then global warming will reach 1.5°C between 2030 and 2050. This forecast is predicated on a global temperature increase at the current rate. The later IPCC Assessment Report 6 (AR6), released in August 2021, also stresses that time is running out for the planet (IPCC, 2021a).

This book can be read as a pilgrimage through an architectural landscape of magical mists, mists of sleep that seem at odds with the science. On waking, a rcal distance always seems to have been traveled; the architectural city – the dream cityof architects – is somehow real, but these dreams are no more real than the circuits of popular culture that feed them. After all, what is a city? And after everything is said and done, we might ask: what are the cities that architects have constructed for themselves and others in the light of the climate emergency? Of course, these cities of architectural dreams have many different scripts and spells. However, these different cities are curiously deaf to the Earth itself and this anacusis prevents architects from healing the planet. It is a deafness that pervades architects and their dreams of the future city. It is a kind of dreaming that appears not to hear the Earth and the ecological violence of modernity.

As I began writing this book, the continent that I live on was aflame as unprecedented mega-fires consumed it. As I finish writing the continent's cities and towns are flooded. The Australian cities clinging to the edge of this continent are surrounded by cultural and ecological landscapes that are too often seen as empty wastelands or fields of extraction. For Australia, as one of the world's largest exporters of fossil fuels, it is ironic that climate change has contributed to the destruction of ecological habitats, ancient cultural landscapes as well as modern infrastructure. Australia's situation, as the driest continent on Earth, with all of its biodiversity, sustained by the world's oldest living culture, points to how settler mentalities, colonialism, and rapacious cycles of economic modernity have failed. In Australia, architecture is embedded in a political economy of fossil fuel production that has been governed in a way that has disregarded the impact of carbon emissions on the planet.

Political Ecology

In hoping to listen to the voice of the Earth, this book constructs a political ecology of architecture. The book examines the human-environment linkages of architecture in relation to power. Political ecology, as a field of knowledge, seeks to ask questions about the relationships between economics, politics, and the natural world (Robbins, 2011). In other words:

> Political ecologists document the power struggles that make and remake "the environment." They provide an understanding of the environment as a dynamic material reality, with exchanges between human and non-human actors, as well as a symbolic

> arena where different (and often clashing) knowledges, desires, and ideologies are cast.
>
> (Kotsila, Paolo De Rosa, & Iengo, 2020)

Any consideration of political ecology regarding architecture must aim to have a focus on the distribution of power. This is because power has shaped the coevolution of nature and architecture. This is different from seeing architecture as a vehicle of urban or technological boosterism or seeing various architectural and spatial practices as solutions to the climate crisis. The above view necessitates interrogating how architects structure power within their discipline through their own theories and practice. Enrique Leff (2021) defines political ecology in the following terms:

> Political ecology is the field where power strategies are conceived, and social struggles are deployed to open new pathways for survival and for building a sustainable future. It involves the deconstruction of modern rationality.
>
> (p. 235)

Political Ecology, emerging in the work of the anthropologist Eric Wolf in the early 1970s, was given further impetus by the work of Blaikie and Brookfield in 1987 (Robbins, 2011; Scheidel et al., 2020, pp. 2, 40). Political Ecology has evolved from being centered on issues in the Global South, to now include examining processes of urbanization in the Global North. As Paul Robbins (2011) notes, political ecology is a field that has attracted "several generations of scholars from the fields of anthropology, forestry, development studies, environmental sociology, environmental history, and geography" (p. 13). Political ecology draws on a range of critical theories, including Marxism and feminist theory, countering Malthusian notions of growth. It should not be surprising that this field has a political dimension since its researchers' advocate fundamental changes in the management of nature and the rights of people. Hence, Political Ecology maps climate issues through "the lens of social, distribution, and knowledge conflicts" (Entitle, n.d).

The reverberations of the above approach will be explored in the following chapters. A range of interwoven narratives regarding architectural discourse, its theories and spatial practices is presented. These narratives will draw inspiration from a broad range of voices: First Nations, schizophrenics, the ecosophy of Felix Guattari, feminist theory, climate justice studies, the UN's Sustainable Development Goals (SDGs), development studies, and, more broadly, the environmental humanities. It will be seen how architects have responded to climate through various modalities and across different contexts. These contexts encompass global cities, technology, architectural science, and the image-politics of social media. The political ecology developed here illuminates how architecture, beyond the limited viewpoints of its own traditions and canon, requires Reworlding – as set out in the final Chapter – to integrate both the human, the nonhuman, and the voice of the Earth itself.

This book draws on architectural history, theory, and the spatial practices of architects themselves. It will be argued that architects are too distantly removed from human-environment systems and ecologies, even as these systems approach points of disruption due to a rapidly heating climate. The voices and signals emanating from these ecological systems are barely heard within architectural discourse. Architects are too enmeshed in their own magic. A new narrative or storying, as some have called it, of architecture through the lens of political ecology offers the hope of leading to a liberatory project. As Kotsila et al. (2020) claim, intrinsic to political ecology are:

> …considerations of justice, equity and fairness in relation to race, gender, class, ethnicity, and other socio-cultural and material inequalities, should be put at the center of research practice and should constitute a shared horizon of values towards collective emancipation.
>
> (Kotsila et al., 2020, Nothing in "nature" section, para. 1)

Another way to describe this book is that in scope, this book is a recent history of architects and the climate emergency since the millennia. The book presents various projects, built and unbuilt, during this time. The text presents and examines the activism of architects as well as exploring the ways they have thought about various policy settings regarding climate change. Inspired and guided by a number of theorists, the book contributes a new perspective regarding the architectural encounter with the Earth's climate. This book builds on and extends my previous book *Architecture as a Global System; Scavengers, Tribes, Warlords and Megafirms* (Raisbeck, 2019). This prior book, through its use of institutional logics, provides a conceptual toolkit for examining how architecture's own professional structures are linked to concatenations of power and knowledge.

Climate Science and the IPCC

Overview of the IPCC Reports

The Intergovernmental Panel on Climate Change (IPCC) reports are the obvious benchmark for assessing current and future issues concerning climate change (IPCC, 2021a). At the time of writing, the current report is Assessment Report 6 (AR6). This has only been partially published: Working Group I *The Physical Science* Basis was published in August 2021, and Working Group II *Impacts, Adaptation, and Vulnerability* was published in February 2022 (IPCC, 2021a, 2022).

The last full IPCC report, Assessment Report 5 (AR5), was published in 2014 (IPCC, 2014). Before that, Assessment Report 4 was published in 2007, and before that, Assessment Report 3 was completed in 2001. Notably, after COP21 in Paris in December 2015, the IPCC was asked to prepare a Special Report (SR15) for publication in 2018 on: the impacts of global warming of 1.5°C above preindustrial levels and related global greenhouse gas emission pathways in the context of strengthening the global response to the threat of climate change, sustainable development, and efforts to eradicate poverty (IPCC, 2018, p. 4).

These successive reports: AR3, AR4, AR5, SR15, and AR6 indicate the emerging scientific consensus on climate over time. The reports serve as a benchmark regarding the science of climate change and its attendant facts. The IPCC reports can thus be seen as reference points from which to assess the dissemination and influence of climate science in architecture.

We might also bear in mind how have the specific technical framings regarding buildings and settlements in the IPCC reports shaped architectural discourse over time. The issues relevant to architecture arising in each Assessment Report are provided in Table 1.1. These issues include: energy emissions from buildings, embodied carbon, Life Cycle Analysis (LCA), Urban Density and Urbanization in the Global South and sea-level rise. At the time of writing, Working Group II and III (AR6 cycle) is to be published in February and March 2022. The AR6 Synthesis report brings together the findings of the Working Groups, and for the AR6 cycle, this report will be published in September 2022. Moreover, the United Nations Framework Convention on Climate Change (UNFCCC) states that the next Assessment cycle (AR7) will have a special report on cities and climate change (IPCC, 2021b). What follows is an overview of the most recent Assessment reports AR5 and AR6, beginning with AR5 (Fig. 1.1).

IPCC AR5

When Working Group III to the Fifth Assessment Report of the IPCC was published, it included a chapter related to Buildings (Chapter 9) (Lucon et al., 2014). In addition, AR5 included a chapter on Human Settlements Infrastructure and Spatial Planning (Chapter 12) (Seto et al., 2014). In the AR5 summary Chapter 9, it was declaimed that:

> ...buildings accounted for 32% of total global final energy use, 19% of energy-related GHG emissions (including electricity-related), approximately one-third of black carbon emissions, and an eighth to a third of F-gases.
>
> (Lucon et al., 2014, p. 675)

The co-authors of Chapter 9 in AR5 stressed how buildings were both critical and necessary to a low carbon future and sustainable development. The AR5 report emphasized building energy use:

> Buildings embody the biggest unmet need for basic energy services, especially in developing countries, while much existing energy use in buildings in developed countries is very wasteful and inefficient.

Table 1.1. Identification of Architecture, Built Environment and Cities Issues in IPCC Assessment Reports.

Report	Date Published	Architecture, Built Environment and Cities Issues Identified
Assessment Report 1	1992	Human Settlements: coastal lowlands and islands; vulnerable populations in developing countries. (IPCC Overview Section 2.4.1)
Assessment Report 2	Synthesis Report October 1995	Scientific-Technical Analyses Of Impacts, Adaptations And Mitigation Of Climate Change — IPCC
Assessment Report 2: Working Group II. Human Settlements:	Synthesis Report October 1995	Sea level rise, most vulnerable human settlemenst. Mitigation: emphasis on technical solutions, renewables and switiching to different energies sources.
Assessment Report 3: Working Group I	March 2001	Options to Limit GHGs: Hundreds of technologies and practices for end-use energy efficiency in buildings.
Assessment Report 3: A Report of Working Group II of the Intergovernmental Panel on Climate Change Impacts, Adaptation, And Vulnerability	March 2001	Human settlements: risk of flooding for many human settlements.
Assessment Report 4	2007	Human Settlements: risk of sea-level rises. Buildings Emissions and solar voltaic technologies, mitigation through building codes.
Assessment Report 5	2014 (note: COP2 Paris in December 2015	Human Settlements: risk of sea-level rises. Buildings Emissions and solar voltaic technologies, mitigation through building codes.

Table 1.1. *(Continued)*

Report	Date Published	Architecture, Built Environment and Cities Issues Identified
Special Report SR15 on Global Warming of 1.5°C.	Published October 2018	Need for: large increases of investments in low-emission infrastructure and buildings.
Working Group I; The Physical Science	August 2021	Cities: intensify human-induced warming locally, urbanization and more frequent hot extremes will increase severity of heatwaves. Urbanization: increases mean and heavy precipitation over and/or downwind of cities resulting in runoff intensity. Coastal cities: more frequent extreme sea level events, extreme rainfall/riverflow events, flooding more probable.
Assessment Report 6. Basis Working Group II: Impacts, Adaptation and Vulnerability	Feb 2022	Human Settlements: risk of sea-level rises. Buildings Emissions and solar voltaic technologies, mitigation through building codes.
Assessment Report 6. Working Group III: Mitigation of Climate Change - IPCC	April 2022	Note: Technical Summary on Urban and Other Settlements and Buildings: "Nearly Zero Energy (NZE) Buildings or low-energy Buildings are achievable in all regions and climate zones for both new and existing buildings." and "There is a significant lock-in risk in all regions given the long lifespans of buildings and the low ambition of building policies."

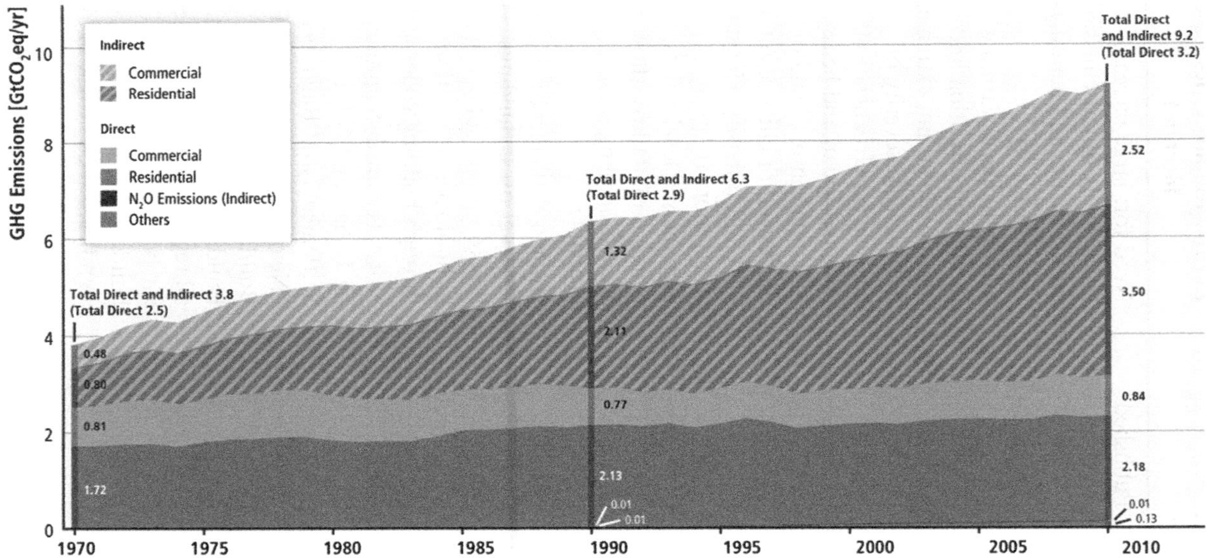

Fig. 1.1. Graph From IPCC, *Climate Change 2014: Mitigation*; Direct and Emissions Indirect (From Electricity and Heat Production) in the Building Subsectors. *Source:* IPCC (2014). AR5 *Climate Change 2014: Mitigation*, Cambridge University Press, p. 678.

> Existing and future buildings will determine a large proportion of global energy demand.
>
> (Lucon et al., 2014, p. 677)

The Chapter contended that best-practice technologies, such as solar hot water or electricity systems, could be implemented in buildings to mitigate energy use:

> In contrast to a doubling or tripling, final energy use may stay constant or even decline by mid-century, as compared to today's levels, if today's cost-effective best practices and technologies are broadly diffused (robust evidence, high agreement). The technology solutions to realise this potential exist and are well demonstrated.
>
> (Lucon et al., 2014, p. 675)

The long-life span of buildings was also seen as a problem because of lock-in risk. Meaning: once a building is built, its carbon emissions will then go into the atmosphere for the lifetime of the building:

> Due to the very long lifespans of buildings and retrofits there is a very significant lock-in risk pointing to the urgency of ambitious and immediate measures (robust evidence, medium agreement).
>
> (Lucon et al., 2014, p. 675)

However, it was noted that:

> Strong barriers hinder the market uptake of these cost-effective opportunities, and large potentials will remain untapped without adequate policies.
>
> (Lucon et al., 2014, p. 675)

At the time of AR5's completion, much of the focus was on energy consumption in buildings rather than the production of embodied energy in their construction (Lucon et al., 2014, p. 694). However, a final section of AR5 Chapter 9 examines issues around embodied energy in buildings, and this was discussed in the following terms:

> However, the embodied energy and carbon in construction materials is especially important in regions with high construction rates, and the availability of affordable low-carbon, low-energy materials that can be part of high-performance buildings determines construction related emissions substantially in rapidly developing countries.
>
> (Lucon et al., 2014, p. 694)

In AR5 Working Group III Chapter 12 of the related AR5 report on Human Settlements, Infrastructures, and Spatial Planning, the co-authors conclude that:

> Thus, a full accounting of GHG emissions from urban infrastructure would need to include both primary and embodied energy of infrastructure materials, as well as energy from the use / operation phase and end-of-life, including reuse and recycling.
>
> (Seto et al., 2014, p. 951)

The accounting of GHG emissions in urban infrastructure and architectural projects are discussed in detail in Chapter 5 Green Architecture and the City.

IPCC Special Report 15 on the Impacts of Global Warming

The *Special Report on Global Warming of 1.5C* was published in October 2018 (IPCC, 2018). The range of risks across the globe identified in SR15 is central to architectural and urban design; these risks include "risks to health, livelihoods, food security, water supply, human security, and economic growth" (IPCC, 2018, p. 9). The IPCC sees an increase in associated risks as global temperatures rise. As reiterated in the report these risks are not uniform and the IPCC argues that some populations are:

> At disproportionately higher risk of adverse consequences with global warming of 1.5°C and beyond include disadvantaged and vulnerable populations, some Indigenous peoples, and local communities dependent on agricultural or coastal livelihoods (high confidence).
>
> (IPCC, 2018, p. 9)

Moreover, the special report identifies regions at higher risk: Especially those with disadvantaged, vulnerable or Indigenous populations and communities dependent on agricultural livelihoods.

> Regions at disproportionately higher risk include Arctic ecosystems, dryland regions, small island developing states, and Least Developed Countries (high confidence).
>
> (IPCC, 2018, p. 9)

The report also emphasized the difference between 1.5°C and 2°C climate impacts.

> Urban heat islands often amplify the impacts of heatwaves in cities (high confidence). Risks from some vector-borne diseases, such as malaria and dengue fever, are projected to increase with warming

> from 1.5°C to 2°C, including potential shifts in their geographic range (high confidence).
>
> (IPCC, 2018, p. 9)

The report emphasized the difference between climate impacts of 1.5°C and 2°C.

> Poverty and disadvantage are expected to increase in some populations as global warming increases; limiting global warming to 1.5°C, compared with 2°C, could reduce the number of people both exposed to climate-related risks and susceptible to poverty by up to several hundred million by 2050 (medium confidence).
>
> (IPCC, 2018, p. 9)

and more worryingly:

> Of 105,000 species studied, 9 6% of insects, 8% of plants, and 4% of vertebrates are projected to lose over half of their climatically determined geographic range for global warming of 1.5C, compared with 18% of insects, 16% of plants, and 8% of vertebrates for global warming of 2°C (medium confidence). Impacts associated with other biodiversity-related risks such as forest fires and the spread of invasive species are lower at 1.5°C compared to 2°C of global warming (high confidence).
>
> (IPCC, 2018, p. 8)

IPCC AR6

As of the beginning of 2022, the IPCC was undertaking work toward the Sixth Assessment Report (AR6) for the sixth assessment cycle (IPCC, 2021a). AR6 comprises three Working Groups: Working Group I is *The Physical Science* basis and brings together the latest understandings of the climate system and climate change (IPCC, 2021a); Working Group II is *Impacts, Adaptation, and Vulnerability* encompassing ecosystems and biodiversity (IPCC, 2022); Working Group III is *Mitigation of Climate Change* assesses progress in mitigation and mitigation options available to different energy and urban systems. This latter Working Group will cover buildings when it is published. When it is published, the *Synthesis Report* will integrate all three Working Group reports as well as the three Special Reports. As discussed above, the *Special Report on Global Warming of 1.5°C* was published in October 2018 (IPCC, 2018). As well, the *Special Report on Climate Change and Land* was published in August 2019 (IPCC, 2019a). *The Special Report on the Ocean and Cryosphere in a Changing Climate* was published in September (IPCC, 2019b).

The AR6 Working Group I report entitled *The Physical Science of climate change* (AR6 cycle) was published in August 2021 (IPCC, 2021a). The AR6 Working Group I had a greater regional focus than AR5 and included a chapter summarizing the science of extreme events. Notably, the AR6 Working Group I report also includes an interactive Atlas (IPCC WGI, 2021). In this report, the IPCC in AR6 notes the Assessment of new datasets and methodological differences between it and the previous 2018 AR5 report. These new methodological advances include:

> New climate model simulations, new analyses, and methods combining multiple lines of evidence lead to improved understanding of human influence on a wider range of climate variables, including weather and climate extremes.
>
> (IPCC, 2021a, p. 5)

Alarmingly, the authors of AR6 note that:

> Global surface temperature will continue to increase until at least the mid-century under all emissions scenarios considered. Global warming of 1.5°C and 2°C will be exceeded during the 21st century unless deep reductions in CO2 and other greenhouse gas emissions occur in the coming decades.
>
> (IPCC, 2021a, p. 17)

For the authors of AR6, perhaps presaging the Special Report on cities, to be published in 2022, cities are seen as central to any understanding of climate change mitigation and adaptation:

> C.2.6 Cities intensify human-induced warming locally, and further urbanisation together with more frequent hot extremes will increase the severity of heatwaves (very high confidence). Urbanisation also increases mean and heavy precipitation over and/or downwind of cities (medium confidence) and resulting runoff intensity (high confidence). In coastal cities, the combination of more frequent extreme sea level events (due to sea level rise and storm surge) and extreme rainfall/riverflow events will make flooding more probable (high confidence).
>
> (IPCC, 2021a, p. 33)

The difference between 1.5°C and 2°C warming is a critical issue for the planet. The degree to which architects are aware of, or motivated by this difference, is an open question. In any case, given the climate science of the IPCC, what does that say about the way that architects create climate imaginaries and deploy spatial practices? Are the good intentions of the architectural imagination enough?

The Conference of the Parties

Alongside the IPCC reports since the millennia, any narrative of architecture's engagement with the climate emergency can also be placed alongside a chronology of the UNFCC's Conference of the Parties (COPs). If the IPCC embodies the science, then the COPs indicate how the science is responded to by states and other actors. The COPs are governed by the United Nations Framework Convention on Climate Change (UNFCCC), the UN entity tasked with supporting the global response to the threat of climate change. The COPs, the Conference of the Parties, are the supreme decision-making body of this Convention. An abbreviated history of the COPs might begin with COP15 and its aftermath. COP15 took place in Copenhagen in December 2009 and perhaps serves as a useful point in time to begin to think about a recent chronology of the politics of climate change. COP15 was significant because it heralded the transition from the Kyoto Protocol Agreement to the development of a new global agreement. This new agreement was then signed in December 2015 in Paris at COP21. At the time of writing the most recent COP was COP26 in Glasgow. As Foyer, Aykut, and Morena (2017) indicate each COP offers "a unique snapshot of the current state of play in global environmental governance, and the attendant – and at times less discernible – spaces of engagement in the climate debate" (p. 2). COP26 is discussed in a little more detail in Chapter 6 in relation to how architects have engaged with environmental politics.

Scope and Sources and Definitions

The panorama of this book is broad and many different voices are given a presence within it. The ideas of different philosophers, theorists and social scientists are referenced; For example, Felix Guattari (Guattari, 1989/2000, 1989/2013), Val Plumwood (1993, 1995, 2012), and the more recent work of anthropologists and political ecologists such as Andrea Nightingale (2018, 2020), Anna Tsing (Feral Atlas, 2021), and Barbara Glowczewski (2020). The work and theories of different architects, both established and emerging, are included in and interwoven across different chapters. The scholarship of many architectural academics – including colleagues at Melbourne School of Design – have been cited and also inspired me as I have written the book. The voices of First Nations and their ancestors also haunt the sentences in this book, as discussed in the final Chapter. The book discusses many of the groups that can be regarded as being a part of architecture's own civil society such as ACAN and The Architecture Lobby. Inspired by Italo Calvino's book *Invisible Cities*, this book presents voices from many different cities, both formal and informal, visible and invisible (1972). Various urban theories and the city are explored and examined. What emerges from this account is more than an exhaustive encyclopedia or taxonomic survey. This is not a survey seeking to describe everything regarding architects and the climate emergency. An encyclopedic account would defeat the purpose of a book that aims to address the urgent issues of the climate emergency in the present. Rather, in the best traditions of architectural criticism, selected architects and their projects are described and evaluated. These architects and projects have been selected to illustrate the most advanced ideas, theories, and practices of architects concerning the climate emergency.

Source Material Within Architectural Discourse

As indicated above, the book traverses and presents a broad landscape of material. In selecting different voices and academic sources, several parameters have been brought to bear on this material. First, the chronological framing of the book is the first 20 years of the millennia. More recent sources in time are given precedence rather than sketching out entire histories. Given the climate emergency and the urgency of addressing this, the book refers to the most recent – and to this author at least – prevailing and advanced narratives in architecture. To reiterate, the material has been chosen to focus on critical moments in climate scholarship or present templates of prevailing or influential views and theories. Architecture is an integrative spatial practice, and consequently, the text has a multi-scalar and transversal focus. Some might call this approach rhizomatic. At times it may seem to the reader there may be an obvious omission or imbalance, for example, the omission of a particular architect, typology, or an overemphasis on one element in the focus of a particular section or Chapter. For example, the book contains very little exposition regarding architectural megafirms and their greenwashing. These firms are discussed in my previous book *Architecture as a Global System* (Raisbeck, 2019). It is hoped that the analytical approaches found here can also be applied by others to particular architects, building types, and various architectural contexts. To address this, it is generally noted where some material is covered and not other material – often further readings are suggested. Finally, the architectural projects discussed in this book are wide-ranging, built, and unbuilt, of different scales and types, and many of those discussed claim to have sustainable or green credentials. Unlike other architectural texts, the survey and analysis presented here does not seek to present and boost exemplary projects; nor does it seek to claim a position for these projects within the Western architectural canon – dare we say the American Ivy League canon. This is book is not a hagiography of anything at all. Overall, the selected architects and projects discussed in each section are intended to illustrate a particular point or contribute to the broader argument that architects have failed to heed the climate emergency.

In architectural discourse, unlike other disciplines, various architectural media and publications, both digital and nondigital, appear to ignore the accepted boundaries between trade magazines and peer-reviewed journals. As Parnell and Sawyer note, architectural magazines are places where architecture is "created," as they argue architectural magazines are important in revealing the current state of play within architectural discourse (2021, p. 44). These magazines have a role in shaping the critical domain of architecture, "incubating" new projects and ideas, as well as being a media where the recent debates of architecture are documented. As the above authors argue, magazines "present and represent architecture" (2021, p. 44). But this situation has also become more complicated with the rise of online magazines and websites such as Dezeen. Arguably, print magazines even in digital form hold more authority than other digital media channels such as Instagram and websites like Dezeen. Nonetheless, as I have described *In Architecture as a Global System*, the "subjugation and commodification of architecture by new media technologies" has continued apace (2019, p. 2).

From within Architectural discourse, architects and projects from the most recent Venice Biennale of Architecture and the Chicago Architecture Biennale have been reviewed and selected. In addition, it should be noted that in architectural discourse, public lectures, particularly those sponsored by the architecture schools, have also been considered to represent key moments in that discourse. These approaches are not to privilege particular viewpoints or foreclose voices associated with less seemingly prestigious schools or venues. The effort has been to describe a fluid and contemporary discourse. Taking all of the above together, readers will find that the book also draws on and cites a wide range of reference material in different media channels including: peer-reviewed papers, conference papers, reports, images and text from websites, podcasts, and online videos such as YouTube. Again, in reviewing, selecting and citing these different sources, the focus here is on those moments and points that appear to this author at least to be the most telling and insightful.

Architectural Jargon and Definitions

How should architects speak about the emerging ecocide? In the Fall 2019 issue of the architectural magazine *Log*, a fetishization of the word carbon is noticeable alongside its derivative architectural terms and titles such as: Global Carbon Form, Carbon re-form, Carbonscapes, Carbon Form, Endgame of Carbon Form, Post-Carbon and Toward Climate Form (Log 47, 2019; AIA Centre for Architecture, 2019). It is almost as if the repetition of this two-syllable word reifies it, granting it a permanent place in an architectural lexicon. The above combinations of syntax, amongst other things, beg the question of what is the right tone for developing a language that speaks to the possibility of extinction? How should the climate emergency be spoken through language? How should the objective facts of climate science be written about? To put the proposition more bluntly: as the anthropogenic climate emergency unfolds with the prospects of mass extinctions, how should this be represented? The self-referentiality of a carbon "language" circulating in architecture may also conflict with and divert us from the reality of climate change. Architects themselves, as it is often anecdotally remarked, have a way of speaking differently to other expert disciplines. Hence, care is taken here to avoid language that might too easily merge into a background of noise, media spin, posttruths – and perhaps even conspiracy theories.

Given the above and to assist the reader, various terms and their problematics are explained where necessary. At various points, and where necessary over the course of this book, the different terms and jargon employed by architects are explained directly. However, this is not a book where the distinctions and definitions of terms are endlessly debated: for example, between sustainable versus regenerative design. The focus is on what architects are currently doing in relation to climate. Of course, many words appear to prevail in the various spheres of architectural climate debate and research: adaptation, mitigation, sustainability, emissions, and triplets like net zero emissions. In the broader discourse of the social sciences, including both architectural and urban theorists, other words abound: Anthropocene, development, sustainability, and ecocide. In this book, many of the above terms associated with different fields of climate discourse are described and used for simplicity sake and for their obvious familiarity to readers.

Chapter Structure of the Book

This book describes many different cities and the voices that speak for these cities. For planners and urban theorists, this approach may seem odd. For many scholars there is perhaps only one, or even only a few kinds of cities: the Global City, cities of the Global South or Global North or cities categorized by geography, planning or population size. But for many architects, Italo Calvino's book *Invisible Cities* is a primary text and this book describes many different types of cities; even though Calvino appears to suggest they are all the same city, Venice (Calvino, 1979). Similarly, all the cities described in this book are also the same city: the supreme Ur city of the Anthropocene. Like Calvino's book, with its narrative of Marco Polo's journeys, the following chapters can be read as journeys across different landscapes. As hinted at in the previous section – and just below – these journeys have abandoned the pretence of an alphabetic survey; the following chapters are not the entire map. Nonetheless, by proceeding in this fashion, a more comprehensive, evocative, critical, and political analysis of how architects have thought about architecture, cities, and the Earth's climate is gained.

Chapter 1 the "Introduction" sets out key background issues and definitions concerning political ecology, climate science and the IPCC, and the outcomes of COP26. The broad scope of material presented in the book is discussed. It is stressed that the book is a critical and selected survey of architecture and climate rather than an impossible to complete encyclopedic survey. The book's scope, use of sources, and selection of architects and architectural projects to be discussed is then outlined. Following an outline of the book's chapter structure, Chapter 1 then proceeds to introduce the broad methodological boundaries of the book. The Chapter concludes with two vignettes of cities, the fictitious Sogo from the 1968 film *Barbarella* and Tuvalu in the South Pacific, indicating the subject range of later discussions in the book. These initial vignettes – coming at the end of Chapter 1 – serve as starting points for the remainder of the book.

The theoretical background that forms the analysis of this book is set out in Chapter Two entitled "Giving Voice to the Peatlands." This Chapter begins with a sketch of the peatland activist BogMan Palm Jaguar aligning him to the post-human ideas of the philosopher Val Plumwood (1993, 1995, 2012). Both these figures point to the need to depart from anthropocentric ways of thinking. Linked to these figures, Felix Guattari's ideas of ecosophy and schizoanalysis are then explored (1989/2000, 1989/2013). As Hanjo Berressem (2020), a cultural studies scholar, argues Guattari's thought is "implicitly ecological" and can also be seen as important to the field of political ecology (p. 3). Guattari's reception and presence in recent architectural theory is then briefly outlined.

Two more theoretical contexts are then discussed in Chapter 2. First, Andrea Nightingale and her co-authors reading of Amitav Gosh's concept of *The Great Derangement* (Gosh 2016; Nightingale et al., 2019). Gosh argues that humankind's inability to face climate catastrophe is a type of derangement. This work leads to considering the biological and physical binary and how, as discussed in Chapter 3, architects try to unite this binary through technology – or technical fixes, as Nightingale and co-authors name it (Nightingale et al., 2019). The final theoretical background in the Chapter focuses on a discussion of specialist expertise and different modes of co-production. Co-production models are important when examining how architects co-produce knowledge with various actors, such as users, clients, and policymakers. In concluding Chapter 2, there is a return to pursuing Guattari and the idea of schizoanalysis.

Chapter 3, "Biophysical Imaginaries" introduces the idea of the biophysical in architecture – as indicated by the chapter's title. The Chapter uses examples of these imaginaries, presenting both built and unbuilt projects, establishing how architects have sought to theorize and graft together the biological and physical. The question asked in this Chapter is how has the idea of the biophysical been imagined in architecture? The Chapter presents the work of Achim Menges and theories of biomimetics. The biophysical is questioned in the projects of the French architect and theorist Francois Roche, active since the millennia, is discussed. Roche's work is compared to the theorization of dark matter and dirt in feminist architectural theory. In order to explore the biophysical in the architectural imagination, a series of floating ecotopias produced by architects since the millennia are described. The work of the London architecture studio Ecologic is then examined as an example of recent efforts to create a new biophysical architecture through the use of algae. The Chapter concludes by asserting that the ecotopian cities of architects, the imaginary melding of the physical and the biological, bear little relationship to the urgencies suggested by climate science.

Chapter 4 entitled "Cities in the Global South" further develops a political ecology of contemporary architecture through a consideration of global cities discourse. The Chapter is framed by the SDGs, and the New Urban Agenda, of UN-Habitat (UN-Habitat, n.d). Following an introductory discussion of the SDGs and sustainability transitions, the Chapter proceeds to present the work of a number of architects. A range of architects and organizations working in the cities and settlements of the Global South are outlined, including: Urban Think Tank, RMA, Studio Mumbai, Yasmin Lari, Communal Taller, Pedro Pacheco (waste and regeneration), Atelier Masomi, and Counterspace. In doing so, various urban theories, community development practices, theories of informality, and decolonization are discussed. The extent to which the presence of the cities of the Global South and climate change are represented in the 2021 Venice Architecture Bienalle. The work of the above architects is then compared to the work of various "star" Architects working in the Global South including,

Norman Foster, OMA, and Bjarke Ingles efforts to masterplan the planet. The Chapter suggests that if embedded in architectural, theory co-production might assist architects in developing new modalities of practice that abandons the tropes of "star" architects. Following this, there is a discussion of SDGs and the NUA; these might be implemented within architectural discourse and practice. Finally, the issue of architectural scale, as a framing device, in urban theory debates is then canvassed. The Chapter concludes for architects' sustainability transitions might best achieved through an engagement with the cities of the Global South.

Chapter 5 "Green Architecture and the Good Anthropocene" examines the idea of the notion of a seemingly benign and imaginary future Anthropocene. This idea is developed through a discussion of a range of architectural projects. How these projects contribute to a notion of a Good Anthropocene is discussed as projects through the lens of social media and associated image-politics across architectural discourse. The 2021 Dezeen Longlist is examined, and selected sustainable projects and their political contradictions are discussed (Dezeen Awards, 2021). The next section entitled "The City of Green Materials" examines how architects use and appropriate green materials, and this is related to issues of embodied carbon. The Chapter then proceeds to look at theories of biophilia in architecture – the biophilic city. Biophilic architectural design and regenerative urbanism are discussed. The Chapter concludes with a section titled "Green Data Cities." This section describes how the proponents of the green data city view the city as a complex system that will bring all the elements of life, both human and nonhuman – in fact of everything – together. By employing various project examples, the section interrogates complex adaptive systems theory and its claims, both indirect and direct, to create low carbon and sustainable cities. Chapter 5 concludes by asserting that architects have constructed their own autopoietic system, a self-reinforcing system of virtue signaling, a system of greenwashed image-politics.

Chapter 6 "Environmental Politics and Design Activism" documents those architects committed to avoiding silence regarding climate and engaged in various modes of design activism. In other words, those architectural groups might be regarded as a part of architecture's own civil society. The Chapter presents the collective and activist activities of architects both at global, multi-lateral, and local levels in relation to climate action: Architects without Frontiers, Architects Declare (UK and AUS), ACAN, World Green Building Council, Insulate Britain as well as Extinction Rebellion. The Architecture Lobby is discussed in relation to the Green New Deal. The activities of the AIA (United States) and RIBA (United Kingdom) are described. In concluding, the chapter discusses the extent and willingness of architects to pursue either nonviolent civil disobedience or violent civil disobedience.

Chapter 7 "Toward a Political Ecology of Architecture" brings together a summary of the above chapters. The Chapter employs a broad interpretation of Malinowski's anthropological study of magic in the Kula exchange system of the Trobriand Islands. To begin with, the Chapter discusses how architectural responses to sustainability are produced and reproduced through a professional culture of autopoiesis. Following this, it is argued that for architects, in regard to the climate emergency, design, itself, is a type of magic. This enables the sustainability responses of architects to be differentiated and described through different categories of sustainability magic: Magic Scripts, Spells and Formulas, Rituals, Shamans and the condition of the performer, Invisibility Magic and Taboos. In this manner, a political ecology of architecture is mapped, as different imaginaries, theoretical positions and practices of architects and architectural theorists are outlined. In conclusion it is contended that the political ecology of architecture can be described as realm of superstition and consequently architects are unable to apprehend the climate emergency or hear the voice of the Earth.

Chapter 8 "Reworlding the Canon" concludes the book. This Chapter begins with a discussion, with the section heading "Other Kinds of Magic," that suggests new theories and analytical instruments that architects might adopt to better address the climate emergency. This section returns to a discussion of Guattari and the prospects of a Schizoanalytic Architecture. What follows is a detailed discussion of the Australian architect Greg Burgess who has worked with numerous Indigenous Language Groups in Australia. His design methods are a template for how architects might better recognize and consider the voice of the Earth in their projects. In answer to this question, elements from the previous chapters are bought together through the concept of Reworlding. Drawing on a range of sources, it is suggested that Reworlding allows architectural discourse to reposition itself. It is concluded that through Reworlding, the climate imaginaries architects can be contested, so they can hear the voice of the Earth as the climate emergency approaches.

Methodological Boundaries

In different ways, the idea of Felix Guattari – himself as a kind of rhizome – haunts this book. Given the somewhat loaded nature of invoking the name of Guattari, a number of contextual points need to be stated at the outset. Guattari's personal biography, their work at the La Borde psychiatric institution, collaboration with Gilles Deleuze, relationship to the events of May 1968, interest in urbanism, political activism, and position as a formative thinker in the emergence of political ecology make their work compelling in relation to architects and the climate emergency. In the first place, Guattari has also been inextricably linked to his sometime collaborator Gilles Deleuze. But as Deleuze and Guattari (1988) themselves state – if indeed they are themselves – at the outset of *A Thousand Plateaus* "Since each of us was several, there was already quite a crowd" (p. 3). As the literary scholar Gerald Bruns (2007) notes in beginning a discussion of the Deleuze and Guattari's concept "Becoming-animal":

> Deleuze and Guattari are notorious improvisers of concepts, which are not always meant to be clear, since for them a concept is never exactly 'about' something, but is a certain way

> of articulating complexities, as if to avoid closure or resolution whatever the matter at hand.
>
> (p. 703)

Guattari was to some extent a creature of the events of May 1968, and gained an aura was amplified in its aftermath, even more so after the publication of *Anti-Oedipus* co-authored with Deleuze originally published in French in 1972 (Deleuze & Guattari, 1983). Guattari was a kind of cult figure associated with 1968. Dagmar Herzog (2016) situates *Anti-Oedipus* as a political and psychoanalytic text rather than a philosophical one. They assert "the book was not just an assault on psychoanalysis. It was also a work of psychoanalysis" (Herzog, 2016, p. 29). Well before this time during the 1950s, Guattari's reputation, and intellectual precociousness, was based not only on his political activism but also their expertise and training in the ideas of Jacques Lacan. Guattari had been one of the first nonpsychiatrists to be invited to Lacan's seminars (Herzog, 2016, p. 38.). Guattari was to become a key figure in the firmament of French intellectuals after 1968 as both doctrinaire Marxism and Freud were critiqued.

The architectural engagement with the Anthropocene is discussed in the following chapters (Steffen, Grinevald, Crutzen, & Mcneill, 2011). As a term, the Anthropocene first came to be used in 2000 by the Noble Laureate and atmospheric chemist Paul Crutzen who was dissatisfied with the existing term the Holocene to describe the current geological epoch. In its place, the Anthropocene was adopted to describe a new geological epoch of the Earth that is entirely dominated by humans. Andrea Nightingale (2018) discusses how the use of the term is not universal:

> The Anthropocene, while widely embraced within political ecology and geography as a signifier of the current epoch of large scale, rapid rate anthropogenic more-than-human transformation, remains a debated concept within the geological sciences wherein the term originated.
>
> (Lövbrand, Mobjörk, & Söder, 2020 as cited in, 2015; Nightingale, 2018, p. 706)

Nightingale draws from "anthropology and development studies on public authority, combined with political ecology insights on resource governance, and feminist theory insights on subjectivity and socionatures" (2018, p. 689). Nightingale is focused on the "socioenvironmental state" and conceptualizes this as a configuration of "contested, shifting, emergent boundaries" that "contain the possibilities for transformative change" (2018, p. 689). Nightingale's effort is to avoid assumptions that see the state as "a (pre)defined actor, hold nature as a separate, interacting domain with society, and propound theories of subject formation that overly emphasize either domination or resistance" (2018, p. 689).

Aligned with Nightingale's methodology, the focus here is on architecture, rather than the state, as a discipline with its own boundary-making processes. This involves questions of how both politics and agency emerge from architecture and its contested, emergent, and shifting relationship to the Earth. In Nightingale's words, it is these "contested, state boundary-making processes that possibilities to reimagine public authority and socio-natures lie" (2018, p. 689). Similarly, this book hopes to avoid the architectural predilection toward static or singularised analytics. Singularized analytics – a singular and sole emphasis on complex systems – offers to architects fixed points for an uncertain future; hence it is important to both identify and avoid fixed boundaries between architecture, the state, society, nature and professional citizenship; what Nightingale designates as subjectivities that are "analytically stable or at least empirically obvious" (2018, p. 689).

Another hagiography attesting to the good intentions of architects, whatever those intentions might be, is not required in the face of the climate emergency. This book rejects the ennui of the contemporary hagiography: the yearning for a saintly style, a sanctified concept, a new design approach, or a new benign future. Two architectural historians, albeit with their own flaws, serve as counterpoints to this insidious tendency. First, Charles Jencks, the preeminent Anglophone historian of the 1970s, thankfully departed from the dreary zeitgeist histories of Zevi, Gideon, and Scully's pandering to the greats. Jenck's histories, drawing on semiotic theory, suggested a complexity and a fluidity of future possibilities as well recognizing, and at times promoting ambiguous readings of the architectural subject (Jencks, 1973). Second, if Jencks appeared to promote ambiguous readings of the architectural canon, then his opposite, if we can call him that, Tafuri rejected it. Tafuri's work gave the historian permission to allow the architectural canon to be read as fragments, yet intimately concerned with power and the way power traverses and is connected to aesthetics (Tafuri & Verrecchia, 1980; Tafuri, 1995). In short, both historians mark the necessity of considering fluidity, ambiguity, fragmentation, and power rather than static and consolidated notions of architecture as a discipline.

This book develops an argument that moves between the interfaces of both architectural theory and practice – this book is not located in one of these domains or the other. Architectural theory and practice are considered here as an archipelago of different ways of knowing and knowledge practices (Frichot, 2018). The degree to which both the theoretical and spatial practices of architects have shifted and are shifting as a result of the climate emergency is described in the following chapters. So many key questions stand out: How have architects theorized the social and the ecological? How is theory co-produced between architects, both practitioners and academics in the space of climate politics? At what points can new formulations of theory emerge? In architecture, where are the new spaces of climate orientated theory, and how have these spaces been narrowed by prevailing threads in architectural discourse (Fig. 1.2).

Conclusion: Sogo and Tuvalu

Sogo was the city featured in the 1968 feature film *Barbarella* directed by Roger Vadim and starring Jane Fonda. The residues of this cinematic city are still present within contemporary architectural imaginaries. Both the architectural critics Charles Jencks and Reyner Banham were captivated by the film's images of

Fig. 1.2. Sogo City, Barbarella (Film), 1968. *Source:* Vadim (1968).

biomorphic forms (Banham, 1981; Jencks, 1971). Jencks linked the set design to the biomorphic and architectural forms of architects then practising in the 1960s. Jencks described *Barbarella* as a film that "gave the first real image of a biomorphic architecture and the triumph of software (Barbarella) over hardware (the excessive machine)" (p. 99). Banham (1981) had also been fascinated by both Barbarella's spaceship and Sogo at the time of the film's release in 1968. Writing in New Society in October 1968, they described the film as the first "post-hardware SF movie of any consequence' arguing that "Barbarella is about responsive environments, of one sort or another and so has been the architectural underground for the last three years or so" (1981, p. 134).

Banham noted that the city of Sogo depicted in the film was a city seemingly at odds with the architecture of Barbarella's spaceship. For Banham, Sogo appeared to be archaic because it didn't "look like anything that could grow naturally out of the inflatable fur-lined technology that dominates the rest of the movie and the software culture within it" (1981, p. 136). He goes on to write, "it is intolerable that a lump of hardware like the city of Sogo could coexist with the living, breathing vision of a friendly, sexy adaptable personal environment" (p. 136). Banham defined responsive environments as those environments which were not "rigid and unyielding; articulated only by hinges between disparate rigid parts; an ambience of curved, pliable continuous breathing and adaptable surfaces" (p. 134). For both Banham and Jencks, the film suggested a new and responsive architecture where environments are shaped by biological processes guided by a new "software" culture that would replace "hardware." Certainly, the above remarks, by Jencks and Banham, suggest a fetishization of technology wedded to a desire for the forms of an adaptable biology.

The architectural engagement with technology in the 1960s and 1970s and its subsequent trajectories in the architecture since that time have echoes in the present. In architecture, as will be seen in the following chapters, these trajectories encompass a range of biophysical and climate imaginaries, including attempts at biophysical engineering, the melding of physical materials and the biological, and the mimesis of biological form to create sympathetic ciphers of nature. Sogo's potential to gratify resided in its twin biological and physical, biotic and abiotic, attributes. Sogo was seen as both an archaic "lump of hardware" and also as a more "sexy friendly, sexy adaptable personal environment" (Banham, 1981, p. 136). Sogo, in these terms, was both hard and functional as well as available and responsive. In the climate imaginaries of architects, these seemingly gendered terms often remain. The present city, as it faces the climate emergency, is too often conceived – in the climate imaginaries of architects – like hardware or infrastructure, as well as a site of flexibility and comfort, a place that can adapt to any external contingency.

The above vignette of Sogo, right at the beginning, serves to contrast the comfort and pleasures of the architectural imagination against an urbanization that is intermingled with various shocks. These are the shocks of catastrophic weather events, geopolitics, corrupt politics, migration, battles for resources, deforestation, fire, drought and desertification and the loss of biodiversity. It is in informal settlements, where urbanization is most problematic, where the real impacts of the climate emergency will play out. As noted by UN-Habitat (2020), these cities have unique vulnerabilities as a result of three factors:

> (i) the physical location is often on fragile sites such as steep slopes, coastlines or floodplains; (ii) the socio-economic characteristics of the residents, such as high levels of poverty and illiteracy mean that these communities have low capacity to deal with climate impacts; and (iii) the political and institutional marginalisation of these communities, stemming from non- recognition of informal settlements as part of the larger city fabric, often results in the absence of meaningful risk-reducing services and infrastructure.
>
> (p. 8)

In this context, and to extend this discussion, we can consider Tuvalu as an island-city in contrast to Sogo. Located in the Pacific, the island of Tuvalu, in Western Polynesia, indicates the immediate need to consider climate shocks, notions of resilience, and migration justice. The government of Tuvalu signed the UNFCCC in 1992, a point in time when it had already recognized its vulnerability to drought, saltwater intrusion, and coastal erosion (Tuvalu, 2007).

This lens is not the only one that needs to be examined in this narrative. Tuvalu is an island state that is at threat because of rising sea levels. Tuvalu is a micro-nation state consisting of an archipelago of low-lying islands especially vulnerable to extreme events and rising sea levels. Climate change will impact Tuvalu's settlements, agriculture, potable water systems, and fisheries. All of these factors are intertwined with the health of the Tuvalu population, including infectious disease transmission. As stated by ESÇAP UN-Habitat (2019):

> Sea level rise will have far-reaching impacts, exacerbating overcrowding, worsening freshwater quality and increasing the risk of infectious disease transmission.
>
> (p. 74)

Tuvalu faces existential risks as a result of being are at the conjunction of climate change and urbanization. At the 2020 UN-Habitat World Urban Forum X (UN-Habitat, 2020a), Katepu Laoi, minister of Local Government and Agriculture of Tuvalu, emphasized the importance of extended family in Tuvalu and how the country was at the front line of climate change (Fig. 1.3).

The above comparison, Sogo alongside Tuvalu, informs a political ecology of architecture. There is a stark contrast between the poetic mists of architects and the reality of rising seas in small nation-states like Tuvalu. If anything else, this

Fig. 1.3. Tuvalu's Minister for Justice, Communication and Foreign Affairs Simon Kofe Speaking in Funafuti, Tuvalu, November 5, 2021. Image Courtesy Tuvalu's Ministry of Justice, Communication and Foreign Affairs. In Acknowledgment of the Work of Tuvalu's Late Minister for Home Affairs & Agriculture Katepu Laoi. *Source:* Tuvalu's Ministry of Justice, Communication and Foreign Affairs/Reuters, (2021, November 9). *Tuvalu minister stands in sea to film COP26 speech to show climate change*. https://www.reuters.com/business/cop/tuvalu-minister-stands-sea-film-cop26-speech-show-climate-change-2021-11-08/.

binary points to the need for architects to develop theories and spatial practices that recognize the intertwining of socio-environmental processes in places subject to the real impacts of disastrous climate change. But perhaps more cogently, these two vignettes indicate a range of issues regarding the presence of the climate emergency in architecture: The desire for, and historiography, of techno-utopian cities; representations of the Global South; urbanization theory as a problematic; exogenous versus endogenous urban instruments and First Nation's knowledge. These issues, as evident in the rest of this book, point to a gap between broader political ecologies of climate change and architectural imaginaries. It will be seen that the residues of Sogo and its trajectories are still ever-present in the architectural imagination. In architecture, this poetic imagination exhibits itself through both its theoretical and aesthetic forms. We can then proceed to ask, alongside the above issues: How do architectural poetics shape architecture's political ecology? How do the poetic dreams, vapors, and magic of architects shape their response to the climate emergency?

References

AIA Centre for Architecture. (2019, September 19). Overcoming carbon form [video]. *Vimeo*. Retrieved from https://vimeo.com/386977949

Banham, R. (1981). In P. Sparke (Ed.), *Design by choice: ideas in architecture*. London: Academy Editions.

Berressem, H. (2020). *Felix Guattari's schizoanalytic ecology*. Edinburgh: Edinburgh University Press.

Bruns, G. L. (2007). Becoming-animal (some simple ways). *New Literary History*, *38*(4), 703–720.

Calvino, I. (1979). *Invisible cities* (W. Weaver, Trans.). London: Pan Books.

Deleuze, G., & Guattari, F. (1983). *Anti-oedipus: Capitalism and schizophrenia* (R. Hurley, M. Seem & H. R. Lane, Trans.). Minneapolis, MN.

Deleuze, G., & Guattari, F. (1988). *A thousand plateaus: Capitalism and schizophrenia* (B. Massumi, Trans.). London: Athlone Press.

Dezeen Awards. (2021). Dezeen Awards, 2021 longlists. Retrieved from https://www.dezeen.com/awards/2021/longlists

Economic and Social Commission for Asia and the Pacific (ESCAP, UN-HABITAT). (2019). *The future of Asian and Pacific cities 2019: Transformative pathways towards sustainable development*. UN Publication. Retrieved from https://www.unescap.org/publications/future-asian-and-pacific-cities-2019-transformative-pathways-towards-sustainable-urban

Entitle. (n.d). The European network for political ecology (ENTiTLE). Retrieved from https://www.politicalecology.eu/

Feral Atlas. (2021). *Feral Atlas: The more-than-human Anthropocene*. Stanford University Press. Retrieved from https://feralatlas.org/

Foyer, J., Aykut, S., & Morena, E. (2017). Introduction. In S. Aykut, J. Foyer, & E. Morena (Eds.), *Globalising the climate: COP21 and the climatisation of global debates* (pp.1–18). London; New York: Routledge.

Frichot, H. (2018). *Creative ecologies: Theorising the practice of architecture*. London; New York: Bloomsbury Publishing.

Ghosh, A. (2016). *The great derangement: Climate change and the unthinkable*. Chicago: The University of Chicago Press.

Glowczewski, B. (2020). *Indigenising anthropology with Guattari and Deleuze*. Edinburgh: Edinburgh University Press.

Guattari, F. (1989/2000). *The three ecologies* (I. Pindar & P. Sutton Trans.). London: Athlone Press.

Guattari, F. (1989/2013). *Schizoanalytic cartographies* (A. Goffery, Trans.). London: Bloomsbury.

Herzog, D. (2016). Desire's politics: Félix Guattari and the renewal of the psychoanalytic left. *Psychoanalysis and History*, *18*(1), 7–37. Retrieved from https://www.euppublishing.com/doi/ref/10.3366/pah.2016.0176

IPCC. (2014). In Core Writing Team, Pachauri, R. K., & Meyer, L. A. (Eds.), *Climate change 2014: Synthesis report. Contribution of working groups I, II and III to the fifth assessment report of the intergovernmental panel on climate change* (p. 151). Geneva: IPCC. Retrieved from https://www.ipcc.ch/report/ar5/syr/

IPCC. (2018). In V. Masson-Delmotte, P. Zhai, H.-O. Pörtner, D. Roberts, J. Skea, P. R. Shukla, … T. Waterfield (Eds.), *Summary for Policymakers. In Global Warming of 1.5 °C. An IPCC Special Report on the impacts of global warming of 1.5 °C above pre-industrial levels and related global greenhouse gas emission pathways, in the context of strengthening the global response to the threat of climate change, sustainable development, and efforts to eradicate poverty* (p. 32). Geneva: World Meteorological Organization. Retrieved from https://www.ipcc.ch/sr15

IPCC. (2019a). IPCC, 2019: Summary for policymakers. In Shukla, P. R., Skea, J., Calvo Buendia, E., Masson-Delmotte, V., Pörtner, H.-O., Roberts, D. C., … Malley, J. (Eds.), *Climate change and Land: An IPCC special report on climate change, desertification, land degradation, sustainable land management, food security, and greenhouse gas fluxes in terrestrial ecosystems*. In press. Retrieved from https://www.ipcc.ch/srccl

IPCC. (2019b). IPCC, 2019: Summary for policymakers. In Pörtner, H.-O., Roberts, D. C., Masson-Delmotte, V., Zhai, P., Tignor, M., Poloczanska, E., … Weyer, N. M. (Eds.), *IPCC special report on the Ocean and Cryosphere in a changing climate*. Retrieved from https://www.ipcc.ch/srocc

IPCC WGI. (2021). *IPCC working group I (WGI): Sixth assessment report. Interactive Atlas*. IPCC. Retrieved from https://interactive-atlas.ipcc.ch

IPCC. (2021a). In V. Masson-Delmotte, P. Zhai, A. Pirani, S. L. Connors, C. Péan, S. Berger, … B. Zhou (Eds.), *Climate change 2021: The physical science basis. Contribution of working group I to the sixth assessment report of the intergovernmental panel on climate change*. Cambridge University Press. In press. Retrieved from https://www.ipcc.ch/report/ar6/wg1

IPCC. (2021b). *Sixth assessment report fact sheet*. IPCC. Retrieved from htttps://www.ipcc.ch/site/assets/uploads/2021/06/Fact_sheet_AR6.pdf

IPCC. (2022). In H.-O. Pörtner, D. C. Roberts, M. Tignor, E. S. Poloczanska, K. Mintenbeck, A. Alegría, … B. Rama (Eds.), *Climate change 2022: Impacts, adaptation, and vulnerability. Contribution of working group II to the sixth assessment report of the intergovernmental Panel on climate change* (p. 3056). Cambridge; New York, NY: Cambridge University Press. doi:10.1017/9781009325844.IPCC

Jencks, C. (1971). *Architecture 2000: Predictions and methods*. London: Praeger.

Jencks, C. (1973). *Modern movements in architecture*. London: Garden City: Anchor Press.

Kotsila, P., Paolo De Rosa, S., & Iengo, I. (2020, November 24). Political ecology. Uneven Earth. Retrieved from http://unevenearth.org/tag/political-ecology

Leff, E. (2021). *Political ecology: Deconstructing capital and territorializing life*. Cham: Springer Nature.

Lövbrand, E., Beck, S., Chilvers, J., Forsyth, T., Hedrén, J., Hulme, M., … Vasileiadou, E. (2015). Who speaks for the future of earth? How critical social science can extend the conversation on the Anthropocene. *Global Environmental Change*, *32*(May), 211–218. doi:10.1016/j.gloenvcha.2015.03.012

Lövbrand, E., Mobjörk, M., & Söder, R. (2020). The anthropocene and the geo-political imagination: Re-writing earth as political space. *Earth System Governance*, *4*(100051). doi:10.1016/j.esg.2020.100051

Log 47: Overcoming Carbon Form. (2019). Anyone Corporation.

Lucon, O., Ürge-Vorsatz, D., Zain Ahmed, A., Akbari, H., Bertoldi, P., Cabeza, L. F., … Vilariño, M. V. (2014). Buildings. In O. Edenhofer, R. Pichs-Madruga, Y. Sokona, E. Farahani, S. Kadner, K. Seyboth, … J. C. Minx (Eds.), *Climate change 2014: Mitigation of climate change. Contribution of working group III to the fifth assessment report of the intergovernmental panel on climate change*. Cambridge; New York, NY: Cambridge University Press. Retrieved from https://www.ipcc.ch/report/ar5/wg3/

Nightingale, A. J., Eriksen, S., Taylor, M., Forsyth, T., Pelling, M., Newsham, A., … Whitfield, S. (2019). Beyond technical fixes: Climate solutions and the great derangement. *Climate & Development*, *12*(4), 343–352.

Nightingale, A. J. (2018). The socioenvironmental state: Political authority, subjects, and transformative socionatural change in an uncertain world. *Environment and Planning E: Nature and Space*, *1*(4), 688–711.

Parnell, S., & Sawyer, M. (2021). In search of architectural magazines. *Architectural Research Quarterly*, *25*(1), 43–54.

Plumwood, V. (1993). *Feminism and the mastery of nature*. London: Routledge.

Plumwood, V. (1995). Human vulnerability and the experience of being prey. *Quadrant*, *39*(3), 29–34.

Plumwood, V. (2012). *The eye of the crocodile*. Canberra: ANU Press.

Raisbeck (2019). *Architecture as a global system; scavengers, tribes. Warlords and megafirms*. Bingley: Emerald Publishing Limited.

Robbins, P. (2011). *Political ecology: A critical introduction*. Chichester; Malden, MA: John Wiley & Sons.

Scheidel, A., Del Bene, D., Liu, J., Navas, G., Mingorría, S., Demaria, F., … Martínez-Alier, J. (2020). Environmental conflicts and defenders: A global overview. *Global Environmental Change*, *63*(102104). doi:10.1016/j.gloenvcha.2020.102104

Seto, K. C., Dhakal, S., Bigio, A., Blanco, H., Delgado, G. C., Dewar, D., … Ramaswami, A. (2014). Human settlements, infrastructure and spatial planning. In O. Edenhofer, R. Pichs-Madruga, Y. Sokona, E. Farahani, S. Kadner, K. Seyboth, … J. C. Minx (Eds.), *Climate change 2014: Mitigation of climate change. Contribution of working group III to the fifth assessment report of the intergovernmental panel on climate change*. Cambridge; New York, NY: Cambridge University Press.

Steffen, W., Grinevald, J., Crutzen, P., & Mcneill, J. (2011/1938). The anthropocene: Conceptual and historical perspectives. *Philosophical Transactions of the Royal*

Society A, *369*, 842–867. Retrieved from https://royalsocietypublishing.org/doi/10.1098/rsta.2010.0327

Tafuri, M., & Verrecchia, G. (1980). *Theories and history of architecture*. London; New York: Granada.

Tafuri, M. (1995). *Venice and the renaissance*. Cambridge, MA: MIT Press.

Tuvalu. (2007). Tuvalu's national adaptation programme of action: Under the auspices of the United Nations Framework convention on climate change. Retrieved from https://unfccc.int/resource/docs/napa/tuv01.pdf

UN-Habitat. (2020). Climate change vulnerability and risk – A guide for community assessments, action planning and implementation. United Nations Human Settlements Programme (UN-Habitat). Retrieved from https://unhabitat.org/sites/default/files/2020/05/climatechange_vulnerabilityandriskguide.pdf

UN-Habitat. (2020a). *The tenth session of world urban Forum, cities of opportunities: Connecting culture and innovation*. United Nations Human Settlements Programme (UN-Habitat). Abu Dhabi, Report, September 2020 [report]. Retrieved from https://unhabitat.org/report-of-the-tenth-session-of-the-world-urban-forum

UN-Habitat. (n.d). New urban Agenda. Retrieved from https://unhabitat.org/about-us/new-urban-agenda

Vadim, R. (Director). (1968). *Barbarella. Marianne productions*. Dino De Laurentiis Cinematografica, Paramount Pictures.

Chapter 2

Giving Voice to the Peatlands

Introduction

Bogman Palmjaguar: A name that serves as a suitable introduction to considering architectural theory and the political ecologies of the climate emergency. Bogman is a useful starting point because his artistic practice, if we can call it that, brings together the perspectives to follow. These perspectives are: the nonhuman evident in the work of Val Plumwood, Guattari's *Three Ecologies*, and political ecology issues in climate and development studies. Bogman Palmjaguar is the name and focus of Luke Fowler's 2007 film and was featured in the Radical Nature exhibition at the Barbican Centre in 2008. On their website Bogman Palmjaguar states:

> —I AM A TRAINED CONSERVATIONIST AND OUTLAWRY OF THE ROCKS CAME TO SYMBOLISE MY ORDEAL. A warm welcome into the world of the rocks hideout. A life at the bottom end of society I could never a afford a car-became an off-road survivor – just as well for me as the 'system' controls the roads and the 'system' is my enemy. RESIST THE 1%-GROSS WEALTH. THE 'SYSTEM' IS ALSO THE ENEMY OF LIFE ON EARTH. And psychiatric horror in the 'system.'
>
> (Bogman PalmJaguar, n.d)

Palmjaguar has spent much of his life, when not incarcerated in psychiatric institutions, wearing eagle and other animal masks. As he is described in Fowler's film, they live in the Flow Country, a peat bog in the far North of Scotland. The Flow Country, as it is called, is an ecological landscape of deep peat dotted with pools of bog. It is a wild habitat and the largest expanse of blanket bog in Europe and has been growing since the last Ice Age. Bogman Palmjaguar is one of the Flow Countries' great advocates for its ecological preservation.

Bogman is a schizo, an exile, a nomad, and an eco-warrior. His life suggests an imbrication with the work of Deleuze and Guattari – particularly Guattari – and Bogman's advocacy for the peatlands suggests how a political ecology of architecture can be approached. In *Anti-Oedipus: Capitalism and Schizophrenia*, Deleuze and Guattari (1983) argue that the schizophrenic is a universal producer.

Architects, Sustainability and the Climate Emergency, 29–50

Published under exclusive licence by Emerald Publishing Limited
doi:10.1108/978-1-80382-291-420221002

Bogman Palmjaguar is just this. A producer who makes no distinction between producing and product. Deleuze and Guattari (1983) employ the example of Antonin Artaud to proclaim:

> The schizo has his own system of co-ordinates for situating himself at his disposal, because first, of all he has at his disposal his very own recording code, which does not coincide with the social code, or coincides with it only in order to parody it. The code of delirium or desire proves to have an extraordinary fluidity. It might be said that the schizophrenic passes from one code to another, then he deliberately scrambles all the codes. (p. 15)

Deleuze and Guattari's collaboration also hints at other concepts and apparitions that we might find in Bogman. Bogman's disconnection from systems of road transport evokes Deleuze and Guattari's (1988) nomad:

> The nomad is not at all the same as the migrant; for the migrant goes principally from one point to another, even if the second point is uncertain, unforeseen or not well localised. But the nomad goes from point to point only as a consequence and as a factual necessity; in principle, points for him are relays along a trajectory. (p. 380)

In a discussion centered on their concept of becoming-animal we might also see in Bogman a figure on the edge of the woods, the image of the shaman or sorcerer:

> Sorcerer's have always held the anomalous position, at the edge of the fields or woods. They haunt the fringes. They are at the borderline of the village, or between villages. The important thing is their affinity with alliance, with the pact, which gives them status opposed to that of filiation. (1988, p. 246)

How might one assess an architect's engagement with the climate emergency from the viewpoint of the schizo, the nomad, the sorcerer, the one becoming-animal? Will the many eyes of Bogman and figures like them help architects to depart from the orthodox subjectivities of their profession? These correspondences and analogies serve to loosen architecture's homogeneity from its own peat-like sediments, layers of history, and its self-reinforcing mirrored ponds. In pursuing the perspective of the schizo, what new sounds or voices might architects hear emanating from ecological landscapes? (Fig. 2.1)

Val Plumwood

Bogman's donning of animal masks corresponds with the consideration of the nonhuman in the work of the Australian feminist philosopher Val Plumwood. In

Fig. 2.1. Bogman Palmjaguar. Image Courstey of Luke Fowlerand LUX Films. *Source:* Fowler, L. (Director). (2008). *Bogman Palmjaguar*. LUX, https://lux.org.uk/work/bogman-Palmjaguar.

her 1995 essay "Being Prey" Val Plumwood (1995) considered the human species not as a predator, but as prey. In describing their encounter with a giant saltwater crocodile, they wrote:

> It is not a minor or inessential feature of our human existence that we are food: juicy, nourishing bodies. Yet, as I looked into the eye of the crocodile, I realised that my planning for this journey upriver had given insufficient attention to this important aspect of human life, to my own vulnerability as an edible, animal being. (2012, p. 10)

In *Feminism and the Mastery* (1993), Plumwood argued that feminist theory is important in articulating theory and discourse across both political ecology and the environmental humanities. They contended that environmental philosophy had been shaped by "masculine presences" that had been given expression through a retained, and all pervading "dualistic dynamic" (1993, p. 2). As a result, Plumwood argued against overdetermined binaries structured by relationships of power. Plumwood's aim was to develop and "increase the critical and analytical force of ecological feminism" in order to "make it a far more powerful political" (1993, p. 1), sought to develop an extended feminist theory that encompassed the environment as well as race, class, and gender (1993, p. 2). Plumwood's project remains today in many guises, and their critique of the critical responses to Cartesian dualism remains relevant today.

Her critique is important in understanding the relationship of architecture to the different nuances of biophysical concepts, as discussed in the following chapters. Plumwood's contribution was to point to the prevailing criticisms of the binary forms of Cartesianism in both empirical and materialist philosophical approaches. Referring to the recovery of the material body in philosophy, they warned that this was not without its dangers. They state of empiricism that:

> Its main import has not been to reinstate the senses and the body, much maligned by rationalism, as sources of knowledge and identity, but to validate a 'scientific', stripped-down, impoverished and subordinated conception of the object of knowledge through observational restrictions on what can count as a knowledge claim and the policing of a strict 'hard'-'soft' subject-object boundary. (1993, p. 121)

And in regard to materialism they argue:

> Materialist positions, which have become popular and self consciously modern positions, attempt to reduce the mental side of the dualism to the bodily, as in physicalism which reduces mind to brain (in mind/brain identity theory), to bodily behaviour (in various forms of behaviourism and stimulus-response theory), or to complex organisational machine states (functionalism). (1993, p. 121)

Going further to caution in her analysis of dualism and its prevailing forms:

> But the original dualism remains in the wings in such a conception to the extent that an impoverished and polarised conception of the material or bodily sphere deriving from the original dualism is affirmed as the ground of reduction. (1993, p. 121)

They identified these diminished approaches as a kind of truncated reversalism. Stating that for these perspectives:

> ...the dilemma of the ghost in the machine is to be resolved by evicting the ghost and embracing its exoskeleton, the empty but serviceable mechanism of the world which is left. (1993, p. 121)

With these cautions in mind, Plumwood's insights allow us to more fully explore an environmental philosophy engaged with the nonhuman world as well as First Nation's relationship to it. Plumwood's (2012) encounter with the crocodile was a life-changing event that led her to reconsider her own philosophical work. This led them to question how western religion and philosophy had conceived of the relationship between humans, animals, and nature, as well as the special status humans had granted to themselves as a species (p. 15). Moreover:

> To hear this voice requires seeing yourself in ecological terms, in historical–evolutionary terms. This crocodile-eye view is the view of an old eye, an appraising and critical eye that potentially judges the quality of human life and finds it wanting. Crocodiles are the voice of the deep past, covering the time span of the rise and extinction of many species. (2012, p. 17)

Writing in the early 1990s Plumwood sought to fashion from an "internal debate" across "green positions," a reshaped political ecology but also a desire for a "liberatory coherence" (1993, p. 14). A coherence within the intellectual and activist groupings of that time alongside coproduced emerging critiques of oppression:

> What is at stake in the internal debate on this issue of political ecology (which has involved social ecology, deep ecology and ecofeminism) is also the question of liberatory coherence and of the relationship between the radical movements and critiques of oppression each of these internal green positions is aligned with. (1993, p. 14).

In highlighting the eye and voice of the crocodile, Plumwood's work is the ground – if we can call it that – for a range of ecological and environmental philosophies that subsequently examined and posited a radical environmentalism.

Guattari, Schizoanalysis, and Architecture

The French psychoanalyst and political activist Felix Guattari is now worth pursuing in more detail. Guattari's thought has been utilized by scholars across a range of social sciences, and in turns, it has been disseminated, studied, and obfuscated. In the field of psychoanalysis, Guattari worked at the La Borde psychiatric institution and collaborated with its founder Jean Oury from 1955 onwards (Dosse, 2007, p. 48). At La Borde Guattari was to state to Oury:

> The central intention is to abrogate various roles and stereotypes: to behave like a madman, a doctor or a nurse, to promote human relationships that no longer lead to lesser roles and stereotypes.
> (Dosse, p. 47)

Through his lifelong engagement with La Borde Guattari developed key concepts in institutional psychotherapy, including the distinction between group and individual phantasms and the concept of transversality (Guattari, 1972/2015). As Deleuze was to note these ideas had "precise practical: orientation introducing a militant political function into the institution, constituting a kind of "monster""; a monster which was not irreducible to particular organisational settings but was "a machine to produce and give voice to desire" (p. 201).

As Simone Brott (2012) describes, after May 1968 and in the early 1970s, Guattari increasingly turned to urbanism. Guattari was intimately involved in the events of May 1968 through his involvement with CERFI Le centre d'etudes, de Recherches, et de formation institutionelles (The Study Center for Institutional Research and Formation). For Guattari, this group was a vehicle for his activism during May 1968 as CERFI morphed into what was called the March 22 Movement. This movement was centered on the occupation of the administration building at the University of Nanterre. In an interview in June of 1968, Guattari was to state of March 22 that CERFI was a "producer of transitional fantasies" and "that it was group that could speak in the mode of free association and constitute itself as the 'analyser' of a considerable mass of students and young workers" (Guattari, 1972/2015, p. 289).

The Three Ecologies

Guattari's book *The Three Ecologies* was published in 1989, written not long after the publication of *Schizoanalytic Cartographies* in 1989 – hence, both books were published in 1989 (Guattari, 1989/2000, 1989/2013). Paul Virilio regarded the former book as the "science of the future" and as such the book is a seminal moment in the history of political ecology (Dosse, 2007, p. 392). Both books outline Guattari's project of understanding the production of subjectivity; "to take note of the reliance of subjectivity on the multiplicity of machines that haunt everyday life and that transform possibilities for better and for worse" (Querrien, 2019, p. 82).

Schizoanalytic Cartographies was described as "Guattari's most inaccessible, sophisticated and logic-based book" (Dosse, 2007, p. 391). The book's publisher Paul Virilio opined that it was a little like Guattari's archives' and "I knew the book would be difficult" (Dosse, 2007, p. 391). Thankfully, *in The Three Ecologies*, Guattari's project of interrogating subjectivities is presented as a framework that can be more easily apprehended. In *The Three Ecologies*, Guattari "defined what he meant by 'ecosophy' as the necessary articulation of between the political and ethical dimension of the environment, social relationships and subjectivity" (Dosse, 2007, p. 391). Typical of Guattari's precociousness, the book has a few uncanny and prescient moments:

> In the field of social ecology, men like Donald Trump are permitted to proliferate freely, like another species of algae, taking over entire districts of New York and Atlantic City; he "redevelops" by raising rents, thereby driving out tens of thousands of poor families, most of whom are condemned to homelessness, becoming the equivalent of the dead fish of environmental ecology.
>
> (Guattari, 1989/2000, p. 43)

If we are to pursue the analogy, the above is a reminder that Guattari's thought, despite all of its complexities – demands for training in philosophy or psychoanalysis – is nonetheless a critical layer in the peatlands of architectural discourse. Guattari summarizes the book's argument and his own hopes for ecosophy:

> A new ecosophy, at once applied and theoretical, ethicopolitical and aesthetic, would have to move away from the old forms of political, religious and associative commitment rather than being a discipline of refolding on interiority, or a simple renewal of earlier forms of 'militancy', it will be a multifaceted movement, deploying agencies [instances] and dispositives that will simultaneously analyse and produce subjectivity. (1989/2000, p. 67)

Central to Guattari's creation of "a new ecosophy" is the exploration and dismantling of the subjectivity configured by what he called "Integrated World Capitalism" and its imbrication with a global mass media. Writing at the conclusion of *The Three Ecologies* he argued that:

> A collective and individual subjectivity that completely exceeds the limits of individualisation, stagnation, identificatory closure, and will instead open itself up on all sides to the socius, but also to the machinic Phylum, to techno-scientific Universes of reference, to aesthetic worlds, as well as to a new 'pre-personal' understanding of time, of the body, of sexuality. A subjectivity of resingularisation that can meet head-on the encounter with the finitude of desire, pain and death. (1989/2000, p. 68)

Guattari completes the book, if such a book can ever be completed, by summarizing his three-pronged polemic: that "escape from the major crisis of our era" can be bought through emphasizing these points with the typography of bullet points rather than prose: "the articulation of a nascent subjectivity" and "a constantly mutating socius" as well as "an environment in the process of being reinvented" (1989/2000, p. 68). It can of course be asked: if the fluid approach adopted here, the effort to produce a mosaic picture of architects and climate, is also too fixed in outlook and method? As Genosko (2009) asks, "why did Guattari settle on a fixed number of ecologies in his book *The Three Ecologies*? How many ecologies are there anyway?" (p. 69).

Hanjo Berressem (2020) argues of all of Guattari's different guises – psychoanalyst, philosopher, activist – then "the most lasting will be the ecologist" (p. 3). Of course, in making this claim Berressem notes that an interest in Guattari is not without its dangers. For example, in his analysis Berressem indicates that *Schizoanalytic Cartographies* is best approached at "a medium level of entry" (p. 53). He states this in order to not risk a dive into a plethora of overwhelming details and the linkage of this detail to Guattari's life and thought and relationship to Deleuze. Cognizant of this caution the aim here is to ask, what are the prospects for what Berressem calls an "architectural schizotheory" (p. 206). To pose the question more expansively, but also with a tone of humility, what might begin to constitute a schizoanalytic architecture and how might this be linked to the Earth's ecologies?

Guattari in Architectural Theory

Given all of their uber coolness – perhaps for architects at least – Guattari's reception within architectural discourse has in the main been sporadic. His overall project, as outlined in *The Three Ecologies*, has been given scant attention in architecture except by a very few architectural theorists as outlined below. Guattari's project of the dismantling of normative patterns of subjectification, the concept of transversality, and their polemic regarding the development of new political ecologies have largely gone unnoticed in wider architectural, theory, discourse, and practice. Even in more recent architectural expositions of Deleuze and Guattari's work, regarding the thematics of affect and infrastructure, Guattari's name is too often relegated to a gloss within citational brackets – or they are simply described as an activist (Jobst & Frichot, 2020). Arguably, it is not unreasonable to say that Guattari has notably been sidelined in favor of his coconspirator, the seemingly more philosophical, Deleuze (Frichot & Loo, 2013). For example, Frichot (2020) expands the politics of affect in relation to architecture's canonical practices and autonomy (p. 13). Likewise, Guattari's thought, perhaps more so than Deleuze's, can also serve as a crack to prise open the regulated disciplinary boundaries between theory and praxis in architecture.

One important contribution to architectural discourse regarding Guattari's thought is seen through the work of Simone Brott. Brott (2016) examines Guattari's engagement with the Japanese architects Shinohara and Takamatsu. As well as this, they also present a close analysis of the genealogy of Guattari's ideas and his relationship to the residual thought of his mentor Lacan. Brott eloquently outlines Guattari's critique of Lacan's psychoanalytic thought, what they call Guattari's "late-Lacanism" finding in it the residues of the Lacanian concepts of "objet petit a (object small 'a') and objet partiel (partial object)." Brott identifies these Lacanian concepts as being linked to Guattari's pursuit of transversality as well as architecture "as the locus for a new image of subjectivity" (p. 97). They go on to write that:

> ...the concepts objet petit a (object small 'a') and objet partiel (partial object), which Guattari used to erect the transversalité. Together, petit a and the object partiel could be said to form the lining of Guattari's entire ethico-aesthetic project. (2016, p. 97)

In tracing and then distilling these residues of Lacan in Guattari's work, Brott suggests that both Guattari's and Lacan's ideas have been "side-stepped" in psychoanalytic thought in relation to processes of subjectivity (2016, p. 97). They intimate that this detour is evident in the Anglo-American reception of French critical theory and thus by implication in contemporary architectural discourse itself. This detour is important to bear in mind, as we track the relationship between architects and the climate emergency, because as Brott bluntly contends:

> For Guattari, subjectivity is not to be found in the detached circuit of repressed signifiers conceived of in psychoanalysis but in the 'external intensities' of cultural, aesthetic, and social reality. (2016, p. 97)

Interestingly, Brott appears to rely on psychoanalytic reading of signifiers as discussed in Chapter 7. Brott, nonetheless – in more detail than Berressem as discussed below – goes on to identify the dissident elements in Guattari's engagement with the Japanese architect Shin Takamatu's syntax building work (2016, p. 99). While more could be written about this, suffice to say, in Brott's exegesis of Guattari's thought in relation to architecture, we begin to glimpse how Guattari's efforts to create emancipatory practices might find an avenue in architecture.

Brott's explication of Guattari is not the only writing that has sought to interpret Guattari in regards to architecture or urbanism. For example, in his earlier work, Gary Genosko establishes how Guattari's posthumanist ideas of urbanism arose from the megamachine concept of the urban theorist Lewis Mumford (Genoskso, 2015). Genoskso points out that Guattari borrowed the concept of the megamachine from Mumford, recoding this concept in order to argue that the city can be seen as an "as a form of machinic enslavement." A form of enslavement that "integrates humans–machines while downplaying the social dimension of subjugation" (p. 7). Genosko (2015) correctly identifies that Guattari's reworking of the "planetary megamachine" was linked to a project that:

> …displaces human subjectivation from the centre of the enquiry and sweeps away the residual humanism of Mumford. Megamachines are pathways to Guattari's post-humanism and animism. (p. 8)

Certainly, Brott and Genosko's interpretations work to untangle Guattari's concepts in relation to architecture and urbanism. However, pursuing Guattari is not without its dangers – as one perhaps needs to repeatedly state. For example, Robert Alexander Gorny (2018) warns against adapting Deleuze and Guattari's concepts reductively in architecture; and perhaps they mean more Deleuze than Guattari in this context. Gorny (2018) explores the notion of assemblage in Deleuze and Guattari's thought and its possibilities within architecture through a lens of materialist orientated theory. In taking up this particular crusade Gorny claims that in architecture "changes do not happen 'in' an environment, but that change takes form as a transformation 'of' a material environment" (p. 189). Going on to contend that "architecture is astoundingly ignorant of its own transformative capacity on a basic level" (p. 189). Despite these somewhat all-embracing claims with their underlying tone of rectitude Gorny (2018) claims a significant point:

> To pose the question "what architecture does," rather than asking "what architecture is" or "what architects do;" focuses the discussion of architecture's agency back onto the ways in which it historically comes to matter as a transformation engine. (p. 189)

A pointed question which in the context of architecture and the climate emergency resounds in later chapters.

To pursue this question in their own manner Gorny examines the intersection of where Foucault and Deleuze and Guattari's intersect, via a reading of the new materialist philosophy of Karen Barad and Rosie Braidotti. Gorny describes this prism of analysis as "an ethological approach to architectural arrangements to overcome an impasse in the understanding of the built environment" (p. 188). In line with the "embodied and embedded, relational and affective" of these thinkers, Gorny concludes that: the new "ecosytemic ethos eventually requires a rigorous reconceptualisation of the built environment as a relational and reciprocal material milieu." (p. 210). For Gorny (2018) the Deleuzo-Guattarian concept of assemblage or *agencement*:

> ...challenges deeply any notion of context or contextualism that incorrectly grounds architectural formations in historical, cultural, expressive, or discursive contexts, yet never within the material arrangements that it transforms. (p. 201)

Citing the cultural theorist Clare Colebrook, Gorny (2018) argues that the city, and indeed architecture, should be examined "not as already differentiated, singular heterotopias, but as further differentiating 'heterogeneses,' which, as Guattari argued, concerns 'processes of continuous singularisation'" (p. 203).

Gareth Abrahams (2019) complements Gorny's work by undertaking a close reading of the concepts of strata and machinic assemblage as set out in Chapter 3 of *A Thousand Plateaus*. He employs these concepts to analyze the different kinds of strata in relation to architectural strata: the building as a physical, organic, and linguistic entity (2019, p. 368). Abrahams approaches the idea of assemblage by pointing to its basis in the linguist Louis Hjemslev's tripartite structure of language used, by Hjemslev, as a theoretical framework to analyze all languages. As he explains, in Hjemslev's thought there were three conceptual planes:

> ...the plane of unformed matter (also termed the 'purport'), the plane of content, and the plane of expression. According to Hjelmslev, a linguistic entity is created within a language when unformed matter transfers across the other two planes.
>
> (Abrahams, 2019, p. 365)

Abrahams points out how Deleuze and Guattari expanded and adapted – arguably in typical fashion – Hjemslev's concept to a "broader corpus" translating and reconfiguring Hjemslev's concept to their own ends (p. 364).

This prompts Abrahams (2019) to examine the territoriality of birds from Chapter 11 "1837: Of the Refrain" in *One Thousand Plateaus*. In doing so Abrahams (2019) attempts to establish how the concept of the strata, a concept integral to the idea of the machinic assemblage, can be employed to understand architectural design and practice. In doing so, Abrahams cautions that "to analyse an assemblage one cannot simply select the variant that best suits the thing produced or study that thing in isolation" (p. 367). Put bluntly, they argue that buildings cannot be simply thought of as physical entities. As they explain it – with reference to *A* Thousand *Plateaus* (p. 73) – the idea of the concept of machinic assemblage goes beyond simply analyzing entities in isolation: such as buildings (p. 367). For example, thinking of a building either as a physical or indeed an organic or biological entity:

> A machinic assemblage is an interstratum insofar as it regulates the relations between strata, as well as the relations between contents and expressions on each stratum, in conformity with the preceding divisions... Finally, the machinic assemblage is a metastratum because it is also in touch with the plane of consistency and necessarily and necessarily effectuates the abstract machine. (2019, p. 368)

In explicating the relationship between strata – arguably, a difficult and an ambiguous concept – and its relationship to machinic assemblage, they strongly argue that a "unified theory of assemblages," as a seemingly obedient concept in urban design and planning theory, is illusory (p. 364). This illusion arises because "many different understandings of this concept can be derived from a detailed examination of Deleuze and Guattari's work" (p. 365). They are thus critical of Delanda's approach and the associated ideas in architecture of autopoiesis limited to physical artifacts and representation alone – as Patrik Schumacher claims. More generally, they bemoan how these concepts have been both defined and employed differently across the spatial disciplines of architecture, planning, and geography; the assemblage concept has taken on different and relativist conceptions; concluding that the various efforts in spatial disciplines to conceptualize the idea of machinic assemblage is often misplaced; reiterating the warning from the Deleuze and Guattari scholar Ian Buchanan (2017) as cited by Abrahams (2019) who:

> ...challenges the idea that the assemblage should be defined and employed without linking it back to Deleuze and Guattari's core texts or their philosophical project more broadly.
>
> (Abrahams, 2019, p. 364)

The intertwined geography of planes of consistency, strata, interstratum, machinic, abstract machine, and assemblage is obviously complex; perhaps, another caution does not bear repeating. Nonetheless, Abrahams poses the question of what might be learnt in thinking of architecture as a "strata/machinic assemblage." As well as asking:

> Can architecture be used to illustrate and explore other chapters of *A Thousand Plateaus*? And across their broader corpus? Where do these illustrations align (and not)? How do these alignments help us piece together different fragments of their corpus? (2019, p. 377)

To return to a previous point, Adrian Parr (2013) underscores Guattari's absence in architectural discourse by tracing the ways in which Deleuze and Guattari's concepts have been employed in architectural theory and practice (p. 197). They argue that the "transformative impulses" of D&G's work has been misdirected because a "largely formalist attitude has remained intact" (2013, p. 197). In surveying the work of the architects Peter Eisenman and Greg Lynn, Parr notes that "architects have chosen to turn a blind eye to the politics" of Deleuze's and Guattari's work. Parr (2013) argues the obvious point "that building is a political practice" and concludes that architects have been amiss in not recognizing the "political potential" of Deleuze and Guattari's thought, stating that: "I remain struck by how little this rocks the architectural establishment" (p. 204).

Schizoanalytic Architecture

To proceed, the above ebbs and flows of Guattari's ideas, their ambiguities of interpretation, and their reception in architecture suggest another tack. Located well outside of architectural theory, Hanjo Berressem's unfolding of Guattari's thought in his book *Felix Guattari's Schizoanalytic Ecology* demands attention (Berressem, 2020). Berressem (2020) concludes that almost all of Guattari's work is either implicitly or explicitly ecological. Berressem argues that architecture is the pivot by which Guattari's thought can be disentangled from Deleuze – and presumably the reception of Guattari through scholars employing Deleuzian lenses. Certainly, Simone Brott's interest in Guattari and architecture suggests this pivot is correct as well (Brott, 2016). It may appear to be disturbing to orthodox architects, but Berressem poses the question of what might constitute schizoanalytic architecture and how might this be linked to the Earths' ecologies and perhaps even its spirit. He argues that Guattari's book *Schizoanalytic Cartographies* is already an "implicitly ecological text," laying out the concepts in his later, more explicit ecological texts, *The Three Ecologies* and *Chaosmosis* (2020, p. 3). Berressem identifies a prescient framing of the ecological in Guattari's oeuvre:

> Although it is only from the Three Ecologies onwards that the ecological comes to provide the explicit frame of his thought, in retrospect, all of his thought can be said to always already have been implicitly ecological. (2020, p. 3)

To restate the above point, Berressem (2020) interprets Guattari's interest in architecture as being "inherently ecological" because architecture was Guattari's "most consistent interest in his work" (pp. 193–194). In linking Guattari's work to both ecology and architecture Berressem pursues the concept of the abstract machine as it "lies at the conceptual centre of Guattari's ecosophic architecture" (p. 194). Going on to argue that:

> Architecture, Guattari maintains should be adequate to the abstract machine which in architectural terms, pertains equally to and paradoxically to both movement and stability as two complementary modes of relating to and inhabiting the spaces of the world. (2020, p. 194)

Berressem argues that these modes should not be considered in binary terms but that "two irreconcilable states are inherent to the overall topology of the abstract machine" (p. 194). These states are what Deleuze and Guattari outline in *A Thousand Plateaus* as the Ecumenon and the Planenom. Berressem explains how Deleuze and Guattari relate the Ecumenon to "settlers, territorialisations, and striation, and the Planomenon to the 'deterretorialized smooth space of nomads.'" For Berressem (2020), these concepts are essential in describing the abstract machine as set out *in A Thousand Plateaus*, which is at the core of Guattari's relationship to architecture and its place and possibilities in the ecologies of the Earth (p. 50). Berressem, after setting out and positioning the idea of the abstract machine and Guattaris' cosmos of thought – what he describes as Guattari's fourfold – they turn to specific architectural issues. Berressem writes:

> In Schizoanalytic Cartographies, the position of the abstract machine as an architectural site is that of an actual rhizome: the abstract site as a deterritorialised field of the possible. (2020, p. 197)

And this establishes that:

> Guattari's ecosophy is keenly aware that there are any number of nonhuman actors at work in ceaseless folding of the given site onto the field of the giving and vice a versa. (2020, pp. 197–198)

In other words, as Berressem so succinctly states:

> It is always within a superposition of the giving and the given that human architecture comes into being. The architectural challenge is to find an adequate way to insert a human habitat into the heterogeneous ecology of the bodies and the spirits of the site; into what the Romans called *genius loci.* (2020, p. 198).

Now, Berressem's case for a schizoanalytic architecture, with an ecology of bodies and spirits, as a given, may seem all too obvious, it nonetheless points to a key question worth restating in Berressem's words: "if the pre- and non-architectural site is described as the architectural unconscious can one envisage a schizoanalytic architecture" (2020, p. 205). Berressem goes on to ask and then answer his own rhetorical question:

> If psychoanalytic architecture is decidedly familial, how would a schizoanalytic architecture – a schizoarchitecture – be more open and communal? Although it would often share its underlying ethos, such an architecture would not exhaust itself in what we tend to think of as green architecture – self sustained, with a low carbon and silicone footprint and energetically up to date – but in the sense of an architecture that is in touch with its ecological unconscious. Along the lines, architectural theory should become a schizoanalysis. (2020, pp. 205–206)

And at this point, we can say that the dream of a schizoanalytic architecture is an architecture that does not simply account for itself as "green" or "low carbon" or "sustainable" or regenerative but is an architecture that is connected to, without limiting to, its own spiritual, nonhuman, and "ecological unconscious" (Berressem, 2020, p. 206). These ideas are further articulated in the final chapter of this book.

Indigenizing Anthropology

In pursuing a schizoanalytic architecture, and such a thing cannot be simply left to rest in the above commentary, we arrive at something quite astonishing. Barbara Glowczewski is an anthropologist who has worked with Warlpiri people in central Australia since 1979. Their work began with their 1981 PhD thesis titled "Le Rêve et la Terre – Rapports au temps et à l'espace des Aborigènes d'Australie" (The Dreaming and the Land – Relations to time and space of Aboriginal Australians). Following this, Glowczewski began a not insubstantial collaboration with Guattari and two seminars were undertaken in 1983 and 1985. These seminars were published as "Warlpiri Dreaming Spaces" under the Part 1 title of "The Indigenous Australian Experience of the Rhizome" in her book *Indigenising Anthropology with Guattari and Deleuze* (2020). Glowczewski links Warlpiri's Yam Dreaming to Deleuze and Guattari's notion of the rhizomatic. Worrying that:

> For those peoples who sustain themselves through gathering yams that grow as underground rhizomes, this term is not a mere poetic metaphor, but a daily reality, a figure of thought that in turn becomes simultaneous a model of social, environmental and mental organisation. This encounter between the form of a plant

> and a way of behaving collectively and thinking, is one among many other ways of interweaving what Guattari called the three ecologies at the core of his ecosophy. (2020, p. 14)

For Warlpiri, the Yam Dreaming is both a space-time and a Law as Glowczewski's notes:

> Where the life of all forms of living beings, be they animals, plants, the rain or the stars, but also with the a multitude of sites that they are connected to and consider sacred (2020, p. 14).

Glowczewski's work may appear to take us a distance from and estrange us from a practical political ecology of architecture. After all what does the Yam Dreaming have to do with architecture? But this feeling of estrangement can only be a sign that Guattari's ecosophy may indeed be valuable. A lack of estrangement, in pursuing new constellations of architectural theory, has more dangers than possibilities. These dangers are threefold. The first appears when theory is oversimplified, prepackaged for distribution to architects literally as a meme; and the second a repositioning, or the cramming, of schizoanalytic thought into the canonical logic of architectural history and theory. Moreover, a lack of estrangement suggests a green architecture stripped of an ecological unconscious. All of these tendencies only serve to reinforce existing orthodoxies of institutional logics within architecture's climate discourse. And thus defeating the purpose of the venture in the first place (Fig. 2.2).

The Great Derangement

The literary scholar Amitav Ghosh alerts us to what they name as the title of their book *The Great Derangement* (Ghosh, 2016). This derangement describes the inability of humans to reimagine a world that is not founded on a separation of the social from the ecological. In late 2019 Ghosh's concept was taken up by 17 coauthors in an issue of *Climate and Development*. In this polemical article Nightingale et al. (2019) use the Great Derangement as an idea to examine why climate change activism and policy has – to put in colloquial terms – hit a brick wall. As they opine:

> Across the globe, scholars, activists and concerned people are decrying the lack of effective action on climate change. Despite decades of warnings over dangerous levels of greenhouse gases in the atmosphere and disturbing variations in ecosystems, emissions have increased rather than decreased. (p. 1)

The above is an all too familiar refrain, in phrases easily caught in a loop-repeat, and architecture is no exception. These coauthors advance their argument by examining how a social-ecological binary promotes the idea that it is

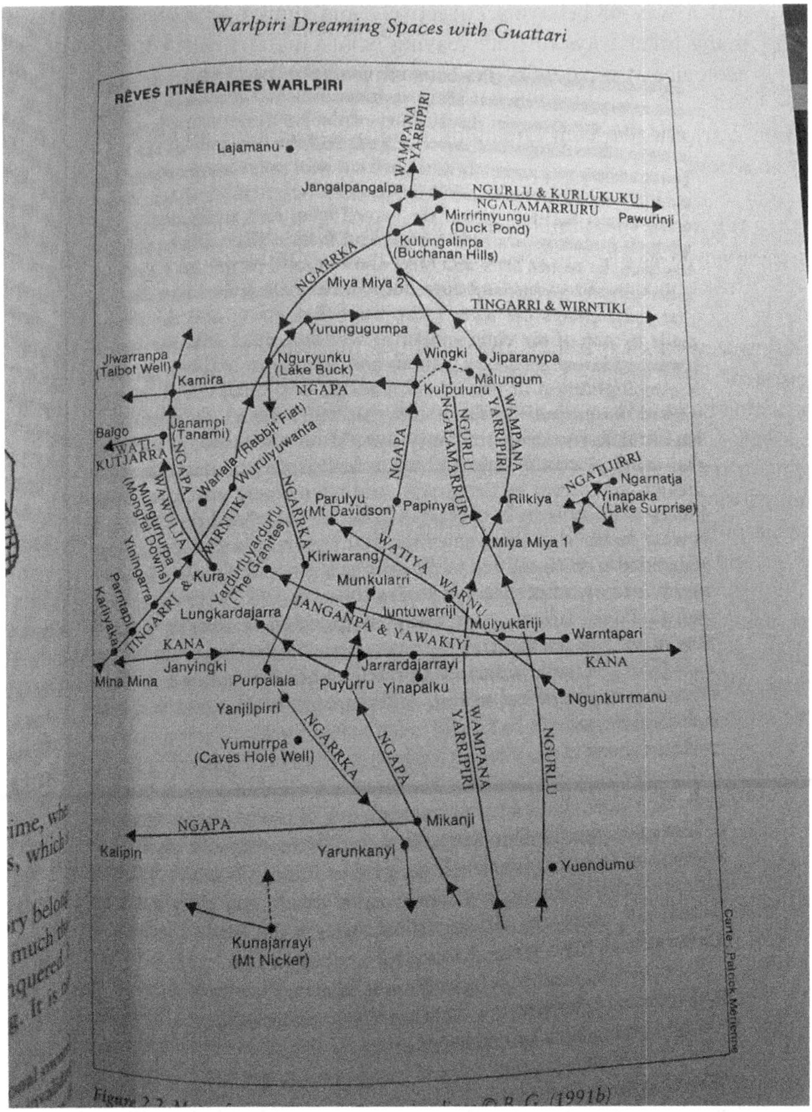

Fig. 2.2. Map of Some Warlpiri Dreaming Lines, 1991. Tanami Desert with Lajamanu to the North and Yuendumu to the South. [606.8 kms by road (13 hours…)] Cartographer Patrick Mérienne. Drawn From Barbara Glowczewski,'s original hand map with meaning of the Warlpiri Dreaming names. The Lines are not roads or pathways just Simplified itineraries of a much more Complex mapping that Warlpiri did not wish to be more precise for public Circulation when First published in B. Glowczewski, *Yapa. Painters from Balgo and Lajamanu* (French/English art Catalog), Baudoin Lebon Editeur, 1991. *Source:* Barbara Glowczewski (2020). Indigenising anthropology with Guattari and Deleuze. Edinburgh University Press. p. 89.

technical solutions, rather than new ways of knowing, that will mitigate carbon emissions in the atmosphere. These solutions, to state it more precisely, bridge the gap between the biophysical Earth, both biotic and abiotic environments, and anthropocentric culture; in the arena of climate development and policy, Nightingale et al. (2019) define the technical in the following terms:

> By technical, we mean responses that impart new infrastructures or that seek to create biophysical changes; examples which vary in emphasis and across scales from water control infrastructures such as irrigation, drinking water and dams, to tree plantations, or geo-engineering for carbon sequestration. (p. 1)

These authors argue the never-ending cycle of "more accurate" assessment of risks:

> ...also, ironically, contributes to limiting the knowledge brought to bear on the climate problem. Limiting knowledge not only shuts down potential solutions, but also generates a variety of social justice concerns. (p. 1)

These coauthors also reflect on the various components of "adaptation and mitigation" policies to highlight the relationship between "a social cartography of vulnerabilities to climate" and the normative remedies "of building adaptive capacity, forging resilience, capturing carbon, and so forth" (p. 2). They conclude that the result of this process is a bundle – dare we say an abstract machine – of processes, produced through a "science policy" interface that is:

> ...absorbed with identifying threats and responding to impacts in order to create transformative change. As a result, the adaptation concept slips insistently to technological measures, despite widespread acknowledgement of their pitfalls.
>
> (Adger, Lorenzoni, & O'Boyd, 2017; Brien, 2009; O'Brien & Selboe, 2015) (p. 2)

In architecture the easy slippage toward technology innovations is made evident in the following chapters. But there is a more substantial issue:

> Truly transformative change – founded on change in knowledge systems and the opening of deliberative space for defining futures – fails to gain traction. (p. 2)

It is important to paraphrase their argument partly in their own words: the focus on the technical creation of new biophysical infrastructure and systems, with boundless functionality, operating across geoscales, makes invisible the "ways of knowing that might help more effectively address the climate predicament and foster determined action" (p. 2) Thus, the binary of the social and the ecological – as well as indicating a kind of derangement – leads to a situation where "moments of concealment are created" (p. 2). In this view, Nightingale et al. (2019) find that climate policy, as practiced through the division of the social and the ecological, has the appearance of an apparent unification of the biological and the technical but has fallen short of: "making politics and social relations pivotal to how system dynamics unfold, and as a result, at best provide a partial entry point to the current task (Brown, 2015; Cannon & Müller-Mahn, 2010; Forsyth, 2018)" (p. 4).

Consequently, they conclude that there is a need to "push the boundaries of socionatural conceptualisations and methodologies to grasp dynamic change" (p. 4). This argument dictates the urgent necessity to examine how biophysical structures, as seen in the next chapters, have been both produced and dreamt of within architecture's own political ecology. The danger is such an examination can easily fall into a descriptive survey celebrating all of the sensory images, deliria, tricks, and sophistries of architects. The biophysicality, and techno-sciences of architectural projects, and associated imaginaries, require a clear-eyed examination that will discern the degree to which the biophysical gesture is a concealing signification.

Mapping Coproduction

Now let us move to the idea of coproduction. This concept is a useful framework for untangling the different disciplinary strands that also constitutes the political ecology of architects, extending the analysis beyond their own disciplinary knowledge. Coproduction enables a framework for examining how architects produce knowledge at a project scale with other actors and entities. In climate-orientated studies coproduction holds out the promise that various boundaries, segmentations, and binary subjectivities can be ameliorated. As Scott Bremer and Simon Meisch (2017) establish, coproduction has emerged "in social science and humanities research on climate change, although there is some ambiguity about the meanings of the term" (p. 1). Through a bibliometric study of coproduction studies, across 130 journals, in climate research Bremer and Meisch (2017) argue:

> We see co-production as a complex meeting place where several different academic traditions and practices converge, overlap, affect each other, come into conflict, or cooperate toward describing and effecting co-production. (p. 1)

Bremer and Meisch (2017) identify two prevalent modes of coproduction in their survey. Firstly, they identify a normative mode that has the purpose of generating "elaborate guidelines (in the most general sense) of how different actors should define and co-produce relevant knowledge" (p. 2). This approach concerns itself with making knowledge; in order for that knowledge to be employed and "demonstrably usable." The second modality of coproduction research – named a descriptive mode – is related to "interpreting the shifting relationships between science, society, and nature – including those around climate change – rather than intervening to actively change these relationships" (p. 2). In this second mode of coproduction research, a more sociomaterial approach is evident, one in which there is an analysis of "various artefacts, settings, knowledge, social orders, and power relationship" (p. 2).

Both of these above modalities appear to be evident in architectural practice but there has been no bibliometric research on coproduction or its modalities in architecture. Evidence for the first mode, the normative mode, relates to the architectural production of sustainability regimes, prescriptions, and codes, as well as making usable knowledge with different communities. The second modality, as described above, may be seen as being linked to the coproduction practices of architects with others across disciplinary fields: urban design, landscape architecture, sustainable design, construction management, engineering manufacturing (e.g., lean construction), ICT, robotics, as well as building science, material, and carbon abatement studies. In this second mode, if we are to follow Bremer and Meisch's (2017) categories, it might suggest that architects also do not actively intervene – preferring to interpret – preexisting project relationships. Hence, for architects, either as researchers or practitioners, coproducing knowledge across these broad disciplines has both theoretical and political implications. We can also ask: Have architectural engagements been overly reliant on positivist and seemingly scientific notions of what science is? More pointedly, if we can discern these two modes in architectural practice, to what extent do either the normative or the overly descriptive engagements of architects mitigate climate change?

Conclusion

Both Guattari's Schizoanalysis and Ghosh's *The Great Derangement* suggest how we might map a political ecology of architecture. If schizoanalysis suggests what might be listened to, then *The Great Derangement* cautions us to be aware of what we may only strain to hear. In addition to these two contexts, coproduction, retrieved for architects from the social sciences, may seem like a more prosaic mode of analysis. However, a consideration of coproduction, in tandem with the above, also serves to outline architecture's political ecology. This is because coproduction is useful in pointing to the way in which architects have coproduced knowledge with others outside of their discipline. Coproduction leads to considering subjectivities of difference, to other voices, to what has been silenced, and to what might be both present and concealed in this contemporary political ecology of architecture.

The approach pursued here, and argued for above, is broadly attuned with Guattari's notion of schizoanalysis. This allows for numerous theoretical perspectives to be drawn upon when discussing the climate emergency. And if this seems both didactic and predictable, via the recourse to a multiplicity of theoretical approaches, it nonetheless still requires emphasis. In pursuing schizoanalysis and drawing on different subjectivities, the dynamic nature of the climate emergency and architecture's relationship to climate science and policy is given due recognition. This more fluid approach acknowledges multiplicities – the identities of the swarm and the horde – and is better able to depict the revolving relationship between architecture and the voice of the Earth. Schizoanalysis enables an exploration of the ecological unconscious of architects. Nowhere is the tone of this voice more evident and more deserving of scrutiny than where architectural theory, research, and practice are intimately intertwined – even more so in the climate imaginaries of architects. If each of these architectural domains has its own vocal range, to what degree do these disciplinary voices respond to a fourth voice – the voice of the Earth:

> Guattari's overriding political concern was with the production of subjectivity, and his green politics were no different. In fact, the problem of the subject's fabrication posed in the context of ecological struggles raises the question of the vast mechanosphere of the globe in the age of advanced informatics which must come to be under human mastery, Guattari claims, if the most pressing problems facing the Earth are to be solved.
>
> (Genosko, 2009, p. 69)

The more important question might be why Guattari's concern only exists in architecture through the barest of hints? At this point it would be premature to speculate as to why this absence exists. The absence of the Earth's voice in architecture is arguably reified by the prevailing emphasis on the subjectivities of political economy and capitalism in architectural discourse – as discussed in Chapter 7 under "Invisibility Magic." Moreover, we can ask to what degree has an emancipatory architectural theory and any corresponding practices related to climate emerged in architecture? Finally, and this third question is more conjectural, in light of the climate emergency, has architectural theory exhausted itself and can it even be redeemed?

The above calls and efforts to enact Guattari's project for the liberation of the subject in architecture cannot be understated. Through Bogman Palmjaguar's scrambling of the codes and Plumwood's ability to see and hear the eye and voice of the crocodile, we come closer to new relationships between architecture, all species, and the voice of the Earth. Yet it must also be considered if the normative relays and ecologies of practice appear to miss so much: the shamanistic, the scatological, the schizophrenic as a universal producer, the solar anus, Antonin Artaud and Nikola Tesla's marriage to the pigeons. If architects and theorists are to pursue a world where the subject is liberated, then this must account for the cosmic and esoteric energies that pulsate and reverberate without imposed distinctions or boundaries.

Guattari's notions of subjectivity point to a bestiary of Animalia – as Gary Genosko (1997) has explored – rather than a strictly human one. Bogman Palmjaguar, Plumwood's saltwater crocodile, and the peat bog with its various pools and layers all point us to reconsidering how existing subjectivities are questioned in relation to the unfolding climate emergency and architecture. How might we give the eye of the crocodile or even the peat itself a voice? If architecture and climate is a peatland to be mapped, what monsters, benign or otherwise, lurk in the peatlands and pools of this topography?

References

Abrahams, G. (2019). The building as a Deleuzoguattarian Strata/Machinic assemblage. *Architectural Theory Review*, *23*(3), 363–379.

Adger, N. W., Lorenzoni, I., & O'Brien, K. L. (Eds.). (2009). *Adapting to climate change: Thresholds, values, governance*. Cambridge: Cambridge University Press.

Berressem, B. H. (2020). *Felix Guattari's Schizoanalytic ecology*. Edinburgh: Edinburgh University Press.

Bogman PalmJaguar. (n.d). Bogman Palmjaguar outlaw rock gardener of 1%. Retrieved from http://www.palmjaguar2.com

Boyd, E. (2017). Holistic thinking beyond technology. *Nature Climate Change*, *7*(97). Retrieved from https://www.nature.com/articles/nclimate3211

Bremer, S., & Meisch, S. (2017). Co-production in climate change research: Reviewing different perspectives. *Wiley Interdisciplinary Reviews: Climate Change*, *8*(482). doi:10.1002/wcc.482

Brott, S. (2012). Collective equipments of power: The road and the city. *Threshold, 2012*(40), 47–54.

Brott, S. (2016). *Architecture for a free subjectivity: Deleuze and Guattari at the horizon of the real*. London: Routledge.

Brown, K. (2015). *Resilience, development and global change*. London: Routledge.

Buchanan, I. (2017). Assemblage theory, or, the future of an illusion. *Deleuze Studies, 11*(3), 457–474.

Cannon, T., & Müller-Mahn, D. (2010). Vulnerability, resilience and development discourses in context of climate change. *Natural Hazards*, *55*(3), 621–635. doi:10.1007/s11069-010-9499-4

Deleuze, G., & Guattari, F. (1983). *Anti-Oedipus: Capitalism and schizophrenia* [R. Hurley, M. Seem & H. R. Lane, Trans.]. Minneapolis, MN: University of Minnesota Press.

Deleuze, G., & Guattari, F. (1988). *A thousand plateaus: Capitalism and Schizophrenia* [B. Massumi, Trans.]. London: Athlone Press.

Dosse, F. (2007). *Gilles Deleuze and Félix Guattari: Intersecting lives*. New York, NY: Columbia University Press.

Forsyth, T. (2018). Is resilience to climate change socially inclusive? Investigating theories of change processes in Myanmar. *World Development*, *111*(C), 13–26. doi: 10.1016/j.worlddev.2018.06.023

Frichot, H. (2020). Infrastructural affects. In M. Jobst & H. Frichot (Eds.), *Architectural affects after Deleuze and Guattari* (pp. 10–25). New York, NY: Routledge.

Frichot, H., & Loo, S. (Eds.). (2013). *Deleuze and architecture*. Edinburgh: Edinburgh University Press.

Genosko, G. (1997). A bestiary of territoriality and expression: Poster fish, bower birds, and spiny lobsters. *Canadian Review of Comparative Literature*, *24*(3).

Genosko, G. (2009). *Félix Guattari: A critical introduction*. London: Pluto Press.
Genosko, G. (2015). Megamachines: From Mumford to Guattari. *Explorations in Media Ecology*, *14*(1–2), 7–20.
Ghosh, A. (2016). *The great derangement: Climate change and the unthinkable*. Chicago, IL: University of Chicago Press.
Glowczewski, B. (1981). *Le Rêve et la Terre – Rapports au temps et à l'espace des Aborigènes d'Australie*. PhD thesis (p. 7). Paris: University Paris.
Glowczewski, B. (2020). *Indigenising anthropology with Guattari and Deleuze*. Edinburgh: Edinburgh University Press.
Gorny, R. A. (2018). Reclaiming what architecture does: Toward an ethology and transformative ethics of material arrangements. *Architectural Theory Review*, *22*(2), 188–209.
Guattari, F. (1972/2015). *Psychoanalysis and transversality: Texts and interviews 1955–1971* [A. Hodges, Trans.]. Los Angeles, CA: Semiotext (e).
Guattari, F. (1989/2000). *The three ecologies* [I. Pindar & P. Sutton Trans.]. New Brunswick, NJ: Athlone Press.
Guattari, F. (1989/2013). *Schizoanalytic cartographies* [A. Geoffery, Trans.]. New York, NY: Bloomsbury.
Jobst, M., & Frichot, H. (Eds.). (2020). *Architectural affects after Deleuze and Guattari*. New York, NY: Routledge.
Nightingale, A. J., Eriksen, S., Taylor, M., Forsyth, T., Pelling, M., Newsham, A., … Whitfield, S. (2019). Beyond technical fixes: Climate solutions and the great derangement. *Climate & Development*, *12*(4), 343–352.
O'Brien, K. L., & Selboe, E. (2015). Climate change as an adaptive challenge. In K. O'Brien & E. Selboe (Eds.), *Adaptive challenge of climate change* (pp. 1–23). New York, NY: Cambridge University Press.
Parr, A. (2013). Politics+ Deleuze+ Guaffari+ architecture. In H. Frichot & S. Loo (Eds.), *Deleuze and architecture* (pp. 197–212). Edinburgh: Edinburgh University Press.
Plumwood, V. (1993). *Feminism and the mastery of nature*. London: Routledge.
Plumwood, V. (1995). Human vulnerability and the experience of being prey. *Quadrant*, *39*(3), 29–34.
Plumwood, V. (2012). *The eye of the crocodile*. Canberra: ANU Press.
Querrien, A. (2019). Schizoanalytic Cartographies. In T. Jellis, J. Gerlach, & J. D. Dewsbury (Eds.), *Why Guattari? A liberation of cartographies, ecologies and politics* (pp. 72–87). London: Routledge.

Chapter 3

Biophysical Imaginaries

Introduction

Once we travel from the realms of theory, the questions then become more specific. How has the idea of biophysical worlds, defined as the biotic and abiotic, been imagined in architecture? In what ways have architects designed projects reconciling physics, the material world of construction and biology. Again, the history of architecture in the 1960s and its relationships to biomorphism shadows this question. For architects during the 1960s, the application of physics to biological systems and the mimicry of physical materials to biological forms was fascinating. Since the early 1960s, biomorphic forms and organic concepts have circulated widely in popular science, popular culture, and architectural discourse. In architecture, in his book on megastructures Justus Dahinden named such schemes "Biostructures" (Dahinden, 1972). Echoing the systems theorist Bertalanffy, he defined this category as the result of the combination of biology with the "science of architectonics" and thus a "synthetic science of living structures" (Dahinden, 1972, p. 38). He saw the proposals of William Katavolos and Rudolf Doernach as attempts to "incorporate the cybernetics of natural processes into man-made structures by programming the building materials to effect subsequent modifications similar to those produced by the chemical processes of dissolution and regeneration" (1972, p. 39). The ubiquity of biomorphism in the 1960s was to prompt Charles Jencks (1971) to proclaim: "when biology becomes the major metaphor of the 1990s, the intuitive tradition will explode in a burst of biomorphic images" (pp. 46–47). Biomorphic formalism was not only the province of groups like Archigram can also be linked to a bundle of neo-futurist tendencies evident in architecture after the CIAM X meeting at Dubrovnik. In December 1958, Yona Friedman established Groupe d'Etudes d'Architecture Mobile (GEAM). GEAM was the first of two groups around which a pantheon of continental (French, German, Swiss, and Dutch) architects assembled (Ragon, 1986, p. 327). By early 1962, the architects reportedly associated with this group included: David Georges Emmerich, Yona Friedman, Gunther Gunschel, Gunter Kuhne, Frei Otto, Paul Maymont, Werner Ruhnai, Eckhard Schultze-Fielitz, Jan Trapman, and Oskar Hansen (FRAC Centre, n.da; Hunter & Harrison, n.d.).

If any one figure – or toy – has come to inhabit this theater of experiment in the 1960s, it is Robbie the Robot. It was Robbie the Robot himself, star of the film

Architects, Sustainability and the Climate Emergency, 51–83

Published under exclusive licence by Emerald Publishing Limited
doi:10.1108/978-1-80382-291-420221003

Hidden Planet, which features in McHale, Hamilton's and Voelcker's exhibit for the *This is Tomorrow* exhibition of 1958 (Whitham, 1990, p. 61). Robbie the Robot is a cypher for the way in which the concept of the biophysical has morphed into a creature, and indeed a city, where nature has been replaced by electro-mechanical creatures. Robbie still anticipates a biological and cultural unification with humans; in his gendered name, in his anthropomorphism and as a kind of collaborative slave to humans. After Robbie, other creatures, machines and monsters have come to represent the sites where physics and biology meet. If Robbie is an early prototype, then Haraway's cyborg is perhaps a more advanced model, somehow a more cultured creature, by becoming a well referenced in critical theory (Haraway, 1991).

And if any reminder is needed about the attributes of neo-futurism, regarding climate change and technical fixes, it is worth noting what Rowe and Koetter (1983) were to remark about early twentieth century Futurism: "the celebration of force is among its more important sustaining sentiments" seeing a parallel between Nietzches the "free man is a warrior" and Futurism's "intrinsic atavism"; an atavism which they also saw evident in the Corbusier's Ville Radieuse minus the nationalism and the "phallic fantasies" (Rowe & Koetter, 1983, p. 30).

Biomimetics

The Urbach Tower

The figure of the robot, chemical processes of dissolution and regeneration, the warrior alongside a certain kind of atavism are evident in the political ecology of architecture today. Materials and the physics of material science are one predominant element in this political ecology. For our purposes, the issues related to the biophysical in architecture can be introduced thorough the Urbach Tower, a folly constructed from self-shaping wood by Achim Menges. Menges has been the founding director of the Institute for Computational Design and Construction (ICD) since 2008 and an influential contributor to the Association for Computer Aided Design in Architecture (ACADIA). Menges (and his collaborator Jan Knippers) argue this tower as "a paradigm shift in timber manufacturing from elaborate and energy-intensive mechanical forming processes that require heavy machinery to a process where the material shapes entirely by itself" (ICD, 2019). The tower itself is 14 m tall and is self-shaping by employing the wood's properties to shrink as moisture content decreases. Menges promotes the tower as highlighting an innovative manufacturing process and an object that intensifies "the visitors spatial involvement and landscape experience by providing a striking landmark building" (ICD, 2019). Menges proclaims that the tower's manufacturing process from forest, to sawmill, and then to machining, and assembly at one location:

> ...allows for sustainable and innovative production, but it also shows, how the self-shaping manufacturing can seamlessly

> integrate in established industrial wood processing and manufacturing workflows.
>
> (ICD, 2019)

Menges' tower is a creature like Robbie the Robot, born from the physics and materiality of organic materials united with manufacturing. If we can allow a different subjectivity, we might suggest that it is Achim Menges himself who is the cyborg: the hand and mind that spans between organic material and robotic processing; the hand – The Hand of the King – that configures the CNC codes and algorithms of timber harvesting and manufacture.

Biomimetic Research

The above outline suggests that in biomimetics a god-like observation of nature can achieve several ends. The central tenet of the biomimetic project is that biology guides construction innovation related to "individual bio-inspired technical features" as well as "methodological strategies" (Knippers, Speck, & Nickel, 2016, p. 3). Biomimetics, and bio-inspired design, aims to explore biological principles through computation enabled by interdisciplinary collaborations (Knippers et al., 2016, p. 5). Architecture and construction have been fertile sites for numerous projects that exemplify biomimetics. In a text that could be a manifesto for the biomimetic movement, Jan Knippers notes how advances in computing allow "natural constructions" which, through the use of algorithms, can for the "first time not only be analyzed but to a certain extent also be transferred to the scale of building construction" (2016, p. 1). But this ambition, the idea of transferring the forms of biology to building construction, then appears to become an exercise in data mining rather than carbon mitigation. Through this logic nature itself is subjected to the processes of data mining – data cleaning, integration, selection, transformation, mining itself, and then knowledge representation: "the exploration and biomimetic transfer of the vast reservoir of nature requires an in-depth analysis and comparison of architecture and biology" (p. 1). This biomimetic transfer takes place through a new interdisciplinary language and is brought into existence through a process of functional identification, extraction, and abstraction:

> Advanced biomimetic research is conceptualised as a dialogue between the disciplines: the functionally important features of the biological systems are abstracted in a model representing the properties of interest and the underlying design principles and structures. (p. 5)

This transfer of abstracted biological codes to building construction also has another impetus: optimization. This is because optimization is seen as an essential part of the evolutionary process: "In the evolutionary process, the efficient usage of natural resources is a significant evolutionary advantage and therefore a guiding evolutionary principle" (pp. 1–2).

The god-like rhetoric of efficiency, and narrative of evolutionary processes, leads to a position that is replete with the elemental structures of physical, chemical, and biological processes. These structures are constituted by "a limited number of chemical elements (often light as to their atomic weight) and basic molecular building blocks" (p. 2). And so these parts, once assembled by evolution, are processed in the following way: these elements become available within a kind of natural laboratory, and are made "directly available," for production, "as part of a natural material loop." This process – suggesting the trope of the laboratory – takes place within a bounded field that is "mostly in the direct surroundings of the organism, and processed under ambient conditions" (p. 2).

But to inhabit this world of biomimetics is to exist in a realm of research and experiment. The outputs of this laboratory are described as being made through "genetically controlled self-organization" that produce "hierarchically organized finely tuned and highly differentiated materials and structures" that feature "multiple networked functions and that sometimes fulfill contradictory functional requirements" (pp. 1–2). And so in this way everything is broken down in the above Lego-like account of evolution; with constituent parts ready for further evolutionary processing. As its proponents Menges and Knippers recognize – but perhaps not the broad mass of architects – this dream of a new era still has a number of obstacles and questions. In the words of Knippers et al. (2016), these questions are threefold: First, how can complex "and adapted biological structures be modeled and simulated?" (p. 4) Second, how can digital computing translate the "design and construction principles found in nature to the macro-scale of building construction?" And the final question appears more alarming, given that the biomimetic project is founded on a type of rhetorical certitude:

> Under which conditions does the abstraction and transfer of biological structures lead to a more sustainable technology that inherits not only the functional, but also the ecologically beneficial properties of the natural role models? (p. 4)

Given the above self-confessed uncertainties of biomimetic processes, even by its own proponents, we might also ask how much does the biomimetic project radically address the climate emergency? In spite of the above questions, the biomimetic process rolls on, a process where biology is examined and data mined; where models are abstracted, designed and simulated and re-simulated; and then up-scaled for implementation into architecture. In this process, nature is seen as a role model – a model seemingly gendered as "mother" nature – that is conflated and grafted on to computing technologies, thus privileging a mechanistic idea of evolutionary biology. Extracting and applying biomimetic shape grammars to advanced construction processes indicates a desire to heal the Earth, a desire to heal the rift between the biotic and the abiotic materials of architecture. Within this earnest desire, as perhaps with all desires, there lurks demons: Foremost amongst these demons are atavistic notions of ecological science, the black box, the architect as the domain controller; controlling a box of actuators of optimized supply chains. Perhaps the worst crime of this biomimetic discourse is the sense of unrelenting ennui that the associated rhetoric gives rise to. This rhetoric makes it difficult to see how the prescriptions and products of this discourse could in any way constitute a liberatory politics.

Ecological Monsters

Biomimetic principles are also evident in the work of the French architect Francois Roche. These appear to be in sharp and, simultaneously, nuanced contrast to the engineered certainties and ennui of the Urbach Tower. Roches' polemical and unbuilt projects, occupying a position in high architectural culture, are a body of work, including exhibitions and films, that appears to pursue a course that employs and explores an oppositional logic. This exploration is a play of oscillation between the biological and the physical; seeking to both intertwine and attack this simple dichotomy through various means.[1] Roche's unbuilt work and theories serve as a useful cypher to narrate the rise of the digital in architecture and the efforts by architects to merge biological notions with digital design technologies. Graduating from the École nationale supérieure d'architecture de Versailles (National School of Architecture Versaille), they set up the studio R&Sie(n) with Stephanie Llavaux in 1989. Roche's work has been extensively exhibited at the Venice Architecture Biennale, and Roche has been affiliated with numerous architecture schools in America and Europe. Roche's identities both as an individual and as an ensemble of studios are fluid. Another incarnation of their studio, named new-territories, was formed in 1993 and is described as "a polymorphous architecture organization" that in its own words "has embraced different labels, names, strategies, and purposes" (new-territories, n.da). Reflecting this, the new-territories studio goes under a variety of names including:

- new-territories
- Laboratory of Digital Disobedience
- R&Sie(n)
- [eIf/bʌt/c]
- M4 (new-territories, n.da)

At the turn of the millennium Roche's work was included in the seminal *Non-Standard Architectures* (2003–2004), exhibition at the Center Pompidou (Migayrou & Mennan, 2003). The trajectories emanating from this exhibition point to how the biodigital has increasingly become a signifier, both directly and perhaps more indirectly, for sustainability and decarbonization in architectural theory and practice. Amongst others, the exhibitors also included Asymptote,

[1]For an overview of films see: https://new-territories.com/props.htm and https://vimeo.com/mindmachinemakingmyths.

Greg Lynn, Kovac Architecture, and UN Studio. The polemics of 1960s architects also shadow this exhibition. Two essays, one by Frederic Migayrpu and one by Zeynep Mennan, reinforce this point (pp. 26–41). The inclusion of two images of biostructures by Rudolf Doernach, the German architect, also points to the 1960s as a historical precursor of the exhibition by its curators and exhibitors. In the exhibition catalog, collated together, are a series of images collated from the history of the twentieth century architecture. These were grouped under the headings of "sequences," "helicoidales," "coques," "formes," and "biomorphismes" (Migayrou & Mennan, 2003, pp. 14–25). The polemic of the exhibition heralded an unashamed return to the geometries of nature and new design technologies; a new techno-organicism, speeded up by software and information systems; a digital organicism that exhibits a plasticity which outstrips Darcy Thompson's diagrams in his book *Growth and Form* (Thompson & Thompson, 1942). A close reading of Roche's unbuilt work, its polemics, and its development since *Non-Standard Architectures* illustrates how the dilemmas of the biophysical is navigated in architectural theory and practice. Four of Roches' projects were presented at the exhibition, and two of these projects, Scrambled Flat 2.0 and [Un]Plug Building, are discussed below (Migayrou & Mennan, 2003, p. 172).

The Scrambled Flat 2.0 Waterflux project has numerous iterations in Roche's oeuvre. The Canadian Center for Architecture (CCA) contains 415 digital files of this project dated between 2002 and 2010 (CCA, n.d). In its initial *Non-Standard Architectures* incarnation, this project is described as a series of farm dwellings bringing together a functional arrangement of hay, wood, men, cows, and bees in each of its arcades. The scheme railed against the European Union's agricultural regulations, which necessitated separating humans and nonhumans. For the community of Évolène, the scheme affirmed their traditional way of living. In the CCA collection, this evolution is described in the following account: "As conceived, Scrambled Flat creates an environment where fluidity between the existence of the animals and the humans is materialised" (CCA, n.d). In a later incarnation, the project evolved into "a geology and glaciology museum and research center focused on the Swiss Alps" (CCA, n.d). This project illustrated the effects of global warming on the Swiss Alps:

> The project was intended to uncover and exorcise the anxieties of ecological disaster, and the principle of flux related to seasonal change and, more broadly, climate change. The firm designed rooms that reproduced the high mountains' geological and meteorological environment, making it visible and experimental, offering refrigerated spaces for art installations and scientific demonstrations.
>
> (CCA, n.d)

The second project, the [Un] Plug Building, plays with the contrast between Euclidean and organic geometries (new-territories, n.db). In this project, this interplay is expressed between a gridded and reactive glass facade onto which organic elements are grafted. These grafted elements, or "melanomas" as Roche names them, appear to erupt from the facade. Roche describes the building in the following terms:

> The façade is thus reactive to this new energy input, making visible the benefits of sustainable energy, but in the same time revealing the pathologies of the sun UV rays exposure, triggering a melanoma of the glass skin...
>
> (new-territories, n.db)

In this project, through naming conventions, unique biological metaphors abound. The facade is described as having a "hairiness" related to its thermal sensors and a "swollenness" of the "facade-curtains" which are described as "productive membranes" related to the face's photoelectric cells (new-territories, n.db). In any case, these seemingly biophysical attributes are achieved through a suggested manipulation of organic forms, grafted compositionally onto a normative office typology. At least Roche appears to understand the difficulties of this oppositional play. Other exhibitors in *Non-Standard Architectures* appear to have abandoned any pretence of doing so, revelling in a biodigital form making that takes those free-forms into the realm of kitsch. With its evocations of flame, emerging from a scrotum suggesting a giant trophy, Kovac's World Trade Center project is a case in point (Architectures nonstandard, 2003, p. 118).

But oppositional logics has its dangers. Writing through a dream like prose text – as the guest editor of *Log 25*, the American architectural journal – Roche's fetishization of binaries leads, perhaps all too easily, to that age-old dilemma within architecture (Roche, 2012). This dilemma is the "violent antagonism" between "two worlds facing each other," the economic versus social forces (Roche, 2012, p. 1). As a result, Roche argues that it is difficult for architects to reconcile:

> On the one hand, technology as a vector of invention in the pursuit of "businessdom," and on the other, the growing of the bottom-up, of the biopolitical tribes, suspicious of the delegation of power's simulacrum.
>
> (Roche, 2012, p. 2)

Tacitly critical of architects such as Rem Koolhaas and other "professionals of the profession," Roche sees these architects as having pursued the roles of the "serious, rigorous, expert, professional" who is essentially "subservient to the system that is supposed to feed them, and in return, logically, they are underpaid for such subservience" (2012, p. 4). Calling this form of practice more sadistic rather than masochistic: "not so much masochism as it is acceptance of the sadism of the system" (2012, p. 4). Roche argues that the way out of this dilemma is to adopt an architectural practice of real masochism, one cognisant of power, exchange, and control. An architecture that might liberate and work through new modes, utilizing from both bottom up and top down a range of new creative practices. And then perhaps echoing a project of schizoanalysis Roche states:

> …as well as multiple crisscrossing, entangled narrations, but also those linked to machinist tools, to the manufacture of things and substances, as well as those navigating this side of that, casting lines of subjectivation, of force, cracks, great escapes, made up of incongruous assemblages, tactical bouts of paranoia and schizophrenia, about to venture into possible new orderings, operative and affective. (2012, p. 5)

Roche's scheme, entitled MMYST, published in *Log 36*, points to how these polemics are melded together with an ecological dynamic as well as advanced methods of architectural computation and construction (Roche & Lacadèe, 2016). Situated in Krabi, Thailand, the project is described as a unique hybrid type, both a farm for birds and a resort for humans. Roche describes the project as being based on "a strategy of mutual exchange between two species – humans and swiftlets – where a human shelter and an artificial bird cave are intertwined" (p. 76). Characteristic of Roche's aesthetic play of binaries, the project is also described through a dialogue between two characters, the mythological greek twins Hypnos and Thanatos. In developing this, Roche argues that the dialogue they use to speak through these figures is a "recurrent debate: scientific idealism versus Anthropocontingencies" (p. 76) (Fig. 3.1).

In parallel with this dialogue, the project traverses the boundaries between the human and the nonhuman, as well as suggesting a process of regeneration through a symbiotic loop between humans and swiftlet birds. The birds make their nests through the excretion of their own saliva. Whereas, for the human's eating the swiftlet's nest in a soup, made from swiftlet saliva it is:

> …believed to help humans maintain a balance energy qi (life energy) and reinforce the immune system. The humans in turn maintain the fragile conditions needed for the bird's reproduction, controlling their indoor habitats temperature, humidity, reverb noises, shadows…
>
> (Roche & Lacadèe, 2016, p. 77)

In the project, this form of mutualism is constructed through robotic means. Through a nozzle, the robotic actuators extrude a "continuum of black, lava like strands, intertwined and knotty" and in the actual prototype this material is described as "starch based bioplastic" (p. 77). In addition, the movement of the nozzle is "directly perturbed in real time by the robots; own noises." and Roche claims – perhaps arrogantly – that this approach allows "continuing the morphology of the existing habitat" (Roche & Lacadèe, 2016, p. 77). The resultant extruded and excreted intestinal sculptures are dead matter, indicating both pleasure and despair; establishing a technical and metaphorical limit of what can be done with robotic extrusions. In the associated dialogue between Hypnos and Thanatos, Roche claims that two different accounts of robotics are in play. On the one hand, there is Hypnos who argues that: "We are instead defining science as a corpus of induction deduction, empiricism, and explanation, dealing with the

Fig. 3.1. New-Territories/M4 and the a Alfred Taubman College of Architecture and Urban Design, mythomaniaS – Concrete[I]Land, University of Michigan, Ann Arbor, Michigan, 2015. *Source:* Architectural Design. Roche, F., & Lacadèe, C. (2016). Parrhesia-stases (The Preamble). Architectural Design, 86(6), 66–71.

necessity of contingencies" (2016, p. 86). And then Thanatos who states: "Is it a crime to be disenchanted, and to develop an aesthetic of this distress, with and within the technologies normally used to wash our souls and treat our minds' affliction?" (2016, p. 86).

Now all of the above may appear to be strangely distant from our aims. However, the conversation, between Hypnos and Thanatos, signals many of the dichotomies identified and explored in this book going forward. At the least, we can say that Roches' use of advanced methods of computing and robotics indicates how these techniques have, more and more in the work of others, been given a place in architecture since the 2003 *Non-Standard Architectures* exhibition. But in the end, all of this play, and wailing, never really reach an escape velocity. Roche is stuck in a mire of digital techniques and robotics, their oscillating binary logic is used to generate a series of reveries, and these reveries, bound to their master's (Roche) actuators, never really embrace ecological voices and the spirits of the Earth: for Roche, it is probably too much fun to chase the scatological – in other words to wallow around in shit – because that is all architects can effect.

To follow this thread, Roche's various fascinations with the scatalogical also have an echo in Janet McGaw's theorization of "dark matter" (McGaw, 2018). McGaw's conceptualization of dark matter is in an effort to explore "architecture's complicity with climate change and possible paths for its redemption" (p. 120). And in order to meet this challenge: "Rethinking materials in architecture must become more than a theoretical practice; It needs to be revolutionary" (p. 121). McGaw defines dark matter as materials produced from carbon, such as crude oil and "speculative practices of resistance" and in her terminology "'dirty' material explorations" (p. 123). These counter explorations of dark matter are related to, amongst others, Guattari's ecosophy. In doing so, McGaw makes a call for us to rethink the architectural relationship to materials arguing that "Matter has agency; Matter matters" (p. 121). Not unlike Roche, McGaw uses the concept of dark matter to call for new speculative practices or as they note, "speculative fabulations," or "art-science worldings," developed in trans-disciplinary practices and design studios (p. 123). All of which is needed to counter "traditional architectural and material practices (p. 127)."

McGaw's position is also evident in the polemics of the feminist architectural theorist Helene Frichot (Frichot, 2020). Frichot rails against the resultant purification that an autonomous subjectivity in architecture engenders. They argue for the joys and liberatory potential of what they call dirty theory. Dirty theory is wary of disciplinary strictures, preferring instead "indisciplination," a wayward approach to problems (p. 7). It is scattergun, if not scatalogical (2020, p. 7). In a close reading of Latour's 2004 essay entitled "Why Has Critique Run Out of Steam? From Matters of Fact to Matters of Concern," Frichot argues that:

> The steam here is a reminder of the socio-technics of thinking, machinic machinations animating technological infrastructures that support and hinder the project of critical theory.... The upshot of this is that critical theory needs better weapons. Dirt bombs to shatter oppressive forces. (2020, p. 59)

For McGaw and Frichot, the allure of dark matter, as with Roche, is an exit for those wearied by normative architectural theory and practice; for those who see more than simplistic binaries in architectural theory; understanding that these binaries are problematic and not invitations for simple solutions. Dark matter's catch cries of art-science worldings, transdisciplinarity, and speculative fabulations potentially expand architectural discourse. However, such creative fabulations, countering the traditional canon, may only lead to a delirium of revolutionary discourse rather than the reworlding of practices in relation to climate. Nonetheless, the discourse of dirt bombs, and renewed subjectivities based on mycelium, excretion, saliva, dirt, and the scatalogical may be the only weapons that architects have in seeking liberatory practices that address the climate emergency. Certainly, the recourse to these subjectivities is preferred to simple-minded and overly earnest biomimetic solutions.

Floating Monsters

A Prelude: Thalassopolis I

But along with the dirt, there is also water, and again, the architectural imaginaries of the 1960s' shadow contemporary impulses. Stimulated by Cold War Technologies and Malthusian concerns regarding food scarcity and overpopulation, architects took to the seas. The oceans became new territories where for the global ambitions of Buckminster Fuller, Yona Friedman, Paul Maymont, Kiyonori Kikutake, Paolo Soleri and Warren Chalk. These architects proposed that marine cities would operate under new regimes of global governance, heralding an era of international cooperation transcending geopolitics and competition between nations (Kaji-O'Grady & Raisbeck, 2005). Whilst the architects of marine cities had global ambitions from the early 1960s onwards, it was not until the 1970s that these ambitions were more fully and subtly articulated. One of the most advanced examples of these projects was Thalassopolis I (1971), designed in collaboration with Edith Vignes and others by the French architect Jacques Rougerie (Rougerie & Vignes, 1978). Thalassopolis I was intended to house various Indigenous People and this marine city was designed as a series of villages on stilts. Thalassopolis I had three primary aims: to ease population pressures in the Indonesian Archipelago, to be a productive mariculture or aquaculture zone, and to foster scientific research (Rougerie, 1972a, 1972b, 1977, 1974; Rougerie & Vignes, 1978). These aims necessitated that the scheme would be directly connected to global population flows and systems of production. To achieve these aims, Rougerie proposed that the scheme use highly sophisticated instruments of localized and international governance to organize the villagers it was intended to house (Rougerie, 1972a, pp. 80–83; Rougerie, 1972b, pp. 34–45) (Fig. 3.2).

The functional structure and governance of Thalassopolis I was described by Rougerie in a number of structural diagrams, "organigrammes" and flow charts. Situated near the Kai Islands, these diagrams indicate that the city was designed for 45,000 people; it was made up of an arrangement of connected floating villages, each with a population of 3,000 people. As described in the diagrams, three of these villages or "unites" would constitute a "Quartiere" of 9,000 people. In turn, the city was then made of five of these quarters and each of these quarters had a specialized administrative or economic function: a central civic quarter, two

Fig. 3.2. Aerial View Thallasopolis I, Jacques Rougerie, 1970. *Source:* Journal of Architecture. Kaji-O'Grady, S., & Raisbeck, P. (2005). Prototype cities in the sea. The Journal of Architecture, 10(4), 443–461.

quarters devoted to fish farming, and one quarter each devoted to crustaceans and algae aquaculture respectively.

In terms of governance, a representative municipal council would provide services to people living in the quarters and villages. An economic directorate as well as an external relations committee would sit alongside this council. An overriding high committee of control – haut comite de control – also with administrative, economic, and financial functions sat above this organizational structure. These layers of governance ensured that Thalassopolis I was connected to international systems of expert knowledge, advice, and control. Over and above this entire structure, Rougerie intended that the entire city would be governed by a unique international convention or treaty rather than a sovereign one (Rougerie, 1972, pp. 80–83; Rougerie, 1972a, pp. 34–45).

Rougerie's scheme, designed for his architecture thesis, under the tutelage of Paul Maymont at UP6 in Paris, thus exhibited a range of intellectual ideas and corresponding compositional instruments. These ideas, as noted above, encompass systems of global governance alongside functional and programmatic determinations. These ideas are then fashioned with the advanced instruments of the time, including diagrams and biomorphic geometries simulating ideas of change and growth. These ideas are arguably related to both Kurokawa's ideas of metabolism and symbiosis, and then diagrammed via the stem and cluster urban and diagrams of Team X (Smithson, 1974). In addition, the scheme aligns itself with the attempts by 1960s architects to develop low-cost housing developed for third world cities, such as the seminal PREVI-Lima housing competition of 1969 (PREVI/Lima, 1970).

Thalassopolis I serves as a prelude, and a historical contrast, to the work of the new marine architects and urbanists, now reacting to climate change. At the core of this contrast is the granular and local-scale detail evident in Rougerie's project with its foci on: localized marine geographies, governance systems, vernacular construction, organic urban form, and acknowledgment of its sea-faring peoples – but a project of colonization nonetheless. The oceans are one test, if not a critical test, of how architectural imaginaries have mapped technology to the realities of biology and ecology. To emphasize the point: building on, and immersing infrastructure in the oceans, is one test of how architects conceive of the interface between the physical and the biological. In the 1960s, architects saw the seas as an escape from the land, an open territory to colonize, urbanize, and extract resources from. Rather than the Malthusian precepts – overpopulation and food scarcity – of the 1960s, architects are now driven by the prospect of melting ice sheets and rising sea levels. This perspective has given rise to a new crop of ecotopian cities on the oceans. In the present, as a result of climate change, the seas are now seen as threatening to engulf lands, low lying cities, and island communities. While this situation appears different, given the existential risks of rising sea levels, as set out in IPCC AR6, we can still ask with Rougerie's project in mind: How much has changed in architectural discourse since the 1960s? What are the socio-technical relationships, design instruments, and technologies invoked and employed, but also absent, in these recent schemes?

Visions of Marine Cities

Conceived in 2001–2002, the Wave Garden project, designed by the Japanese architect Yusuke Obuchi, is a floating island located off the San Luis Obispo coastline of California. Notably, Obuchi was Co-Director of the Design Research Laboratory at the Architectural Association (AA) between 2005 and 2010. The project was conceived of as an artificial island constructed from what is described as 1,800 tiles comprising a three-inch thick "piezo-electric membrane" (Beekmans, 2012; Pruned, 2005). Through wave action, these tiles generate electricity as these oscillate with the waves creating an undulating landscape. Uniquely, the Wave Garden is a futurist project that accounts for the energy consumption habits of humans as well as being a landscape that can be visited on weekends for recreation. However, these recreational visits can only happen if the population of the adjoining coastline have consumed minimal energy during the preceding weekdays. When this happens, the coastal population "are rewarded: the platform rises and forms recreational spaces and swimming ponds. But if weekday demand is too high, the platform remains strictly a power plant" (Beekmans, 2012). Through this regime, the entire project thus operates as a "barometer of energy use" (Beekmans, 2012). Thus directly enabling unseen power generation to be made visible and linked to the energy habits of the adjoining coastal population. In an interview, when questioned about the viability of floating architecture and infrastructure Obuchi was to state: "It is totally reality. It is already happening pretty much everywhere around the world" (Beekmans, 2012).

Certainly, by around this time, along with the work of the IPCC's, Third Assessment Report published in 2001, data-driven instruments of simulation and representation gave the 1960s marine ecotopias a new lease of life in the architectural imagination. Widely disseminated in 2005 amongst architects was Michael Batty's work at the Center for Advanced Spatial Analysis, University College London. Batty used GIS technologies to simulate the flooding of London as a result of climate change; the simulation described what would happen if sea levels rose by 10 m if the Greenland ice cap melted (Batty & Hudson-Smith, 2005) (Fig. 3.3).

If we are to contrast Rougerie's and Obuchi's schemes with something more recent, less earnest, and arguably deferential to fossil fuel extraction, then it might be the underwater skyscraper or seascraper, the size of the Empire State Building, designed by Sarly Adre Sarkum of Malaysia and exhibited at the 2010 Venice Architecture Biennale. Sarkum went on to design and curate the Malaysian Pavilion exhibits at Venice in 2012 and 2014. The project was presented as a miniature model of the seascraper sitting in a small glass cylinder or vase filled with clear jelly to simulate the water. The proposal is 150 m below the water and provides its inhabitants with apartments designed in a minimalist aesthetic

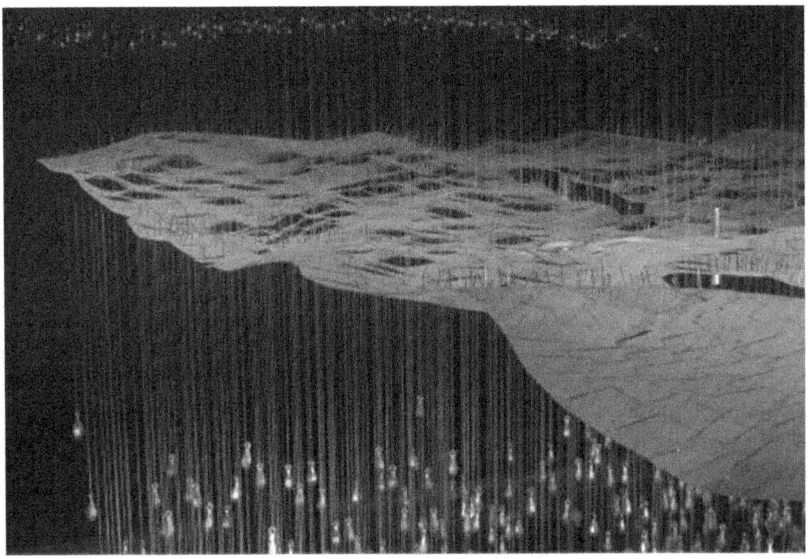

Fig. 3.3. Wave Garden, Yusuke Obuchi, Princeton University's School of Architecture, 2002. *Source:* Image Yusuke Obuchi. Architectural Design. Berman, I. (2010), Amphibious Territories. Archit Design, 80: 66–73. https://doi.org/10.1002/ad.1077.

looking into the ocean. The concept was predicated on the notion that "rather than tethering our buildings to the sinking shoreline why not suit them for a life at sea?" (Yoneda, 2011). This is a future where climate change has run its course as the architect proclaims:

> ...a future where land as a resource will be scarce; it is only natural progression that we create our own. Approximately 71% of the Earth's surface is ocean, even more if climate change has its way, hence it is only natural progression that we will populate the seas someday.
>
> (Yoneda, 2011)

Floating iceberg like in the sea, and circular in plan, the portion of the water-scraper above the water included a forest, uncannily reminiscent of the forest in the science fiction film *Silent Running* (Gruskoff & Trumball, 1971). Not unlike the space ship in that film the seascraper is cut adrift: utilizing wave, wind and solar power and produces its own food through, farming, aquaculture and hydroponics. Cephalodic tentacles swirl around the lower reaches of this city. As with other marine cities the biomorphism of cephalopods – aka jellyfish – are invoked and these "glowing tentacles provide sea fauna a place to live and congregate while collecting energy through its kinetic movements" (Fox, 2010). Sarly is president of the Malaysian Green Building Council, as well as having clients such as Petronas, and Dyna who are consultants to the oil and gas industry ([SA]Squared, n.d.). At the 2010 Biennale the project sat beside other projects in the exhibition, including the FELCRA tower, a proposal celebrating the palm oil industry in Malaysia.

Shizimu Ecotopia at Sea

Perhaps inspired by the floating cities of the Japanese Metabolists, infrastructure engineering wedded to a rhetoric of organic forms and processes are also evident in the 2008 Shizimu corporation's Green Float project (Shizimu, n.d; wpengine/Seasteading Institute, 2010). Shizimu has expertise in construction, architecture, civil engineering, and property services and has annual revenues of over $1B USD. Shizimu claimed the concept would be carbon negative, generate no waste and provide food self-sufficiency for its inhabitants. Moreover, it was claimed that this city would help to clean up and *purify* the Great Pacific Garbage Patch by capturing and using garbage in the city as an energy source. Conceived as a lily pad with a central 1km high tower, the cities' design and aesthetics was structured through biomorphic metaphors. Each lily pad like city of 10,000 to 50,000 people was seen as a component that could be linked to other similar pads and in this fashion be equated to organic growth. The technologies proposed for the city include renewable energy and "CO2 Recovery and Ocean Sequestration" as well as "a mixture of forest, fields, waterways, reservoirs and grasslands" (Shizimu, n.d). A recycling plant would help convert waste into energy, and because the city is located on the equatorial oceans, this would enable large quantities of food to be grown through aquaculture. Water purification would also be an element within the schemes bundle of technologies.

Unexpectedly, these schemes never account for geopolitics – are they defensible structures? Not unexpectedly, these types of proposals often include all the accoutrements of a neoliberal fantasy. The Shizimu scheme included a resort element that featured townhouses surrounded by built-in lagoons and summer beaches (Meinhold, 2010a). In a similar vein, the 2010 project the Gyre designed by the Canadian firm Zigloo included an off-shore resort complete with retail shopping and was described as an "underwater eco-skyscraper" (Meinhold, 2010b). This project was intended to be a "floating eco-development meant to be both a research station" and a "green design and lifestyle site": a project that might be needed for rising sea levels where humans "end up living in a water world" (Meinhold, 2010b).

Not all of these floating proposals seem so ignorantly embedded in global capitalism or mercantile economics. At the other extreme, and amongst all the above, there are also projects such as The Venus Project. The Venus Project is a global group founded by the Futurist Jacque Fresco, whose aims are to produce through social action a "peaceful and sustainable global civilization" and where "human rights are no longer paper proclamations but a way of life" (The Venus Project, n.d). Facilitating this global vision, The Venus Project has proposed that floating cities are needed: which might "easily accommodate many millions of people and relieve the land-based population pressures" (Venus Project, n.d). Such cities would, according to the project's adherents, provide oceanographic education, aquaculture, and deep-sea mining in a way that would not disturb the ecology of the oceans (The Venus Project, n.d).

Another proponent since the millennium of floating cities is the Belgian born Paris based architect Vincent Callebaut. Callebaut's beautiful Lillypad is a design for a floating arcology reminiscent of the Arcologies of the Italo-American architect Paolo Soleri. Each Lillypad would allow 50,000 people to live at sea. Swarms of these pads would form ecotopias entirely at sea, capable of sustaining themselves without undesirable environmental impact. This project is not without its altruism, also being intended as a city for climate refugees (Vincent Callebaut, Architectures n.da). Another Callebaut project is Aequora an Oceanscraper intended to be "Printed in 3D from the Seventh Continent's Garbage" (Vincent Callebaut Architectures, 2015). The project was intended to travel the five main ocean currents, or gyres, off the coast of Rio De Janeiro Brazil. Even though its underwater section draws on the helical forms of shells, the project's overwhelming metaphor is the jellyfish: "Bio-inspired, the farm draws its name from a bioluminescent, light-emitting jellyfish characterised by its articulated, webbed tentacles' (Vincent Callebaut Architectures, 2015). Not surprisingly, the tropes of resort architecture, with sustainability credentials, shadow, these marine projects. In 2017 Callebaut produced a scheme for the Nautilis Eco-Resort, a Biomimetic Learning Center located in Palawan the Philippines, and then claimed this project as being "Zero-Emission, Zero-Waste, Zero-Poverty" (Retalk Asia, n.d). Not dissimilar to the earlier Thalassopolis I, this project is described in the following terms:

> ...a pioneering, eco-tourism complex designed to unite the scientific community with eco-tourists. Based on "biomimicry," the architecture of this Eco-Resort is inspired by the shapes and structures that exist in living beings and endemic ecosystems. By minimising its ecological footprint, the experience is centred on the preservation of nature and local urban ecology while respecting endemic ecosystems and agro-ecosystems.
>
> (Retalk Asia, n.d)

Callebaut describes himself as an Archibiotect. This terminology is a combination of prefixes or perhaps suffixes – as if to reinforce the credential of an already prolific practice – as a consequence of Callebaut's own interests in "ARCHItecture + BIOtechnologies + Technologies of Information and Communication" (Vincent Callebaut, Architectures n.dc).

Callebaut has won numerous awards, including the Global Human Settlements Award on Planning and Design (SCAHSA) in 2017. This award is sponsored by the not for profit Global Forum on Human Settlements (GFHS) (Vincent Callebaut Architectures n.db) (Fig. 3.4).

The different urban scales of these floating visions are not always so large, with their Buckminster Fuller like desires to address global problems. Architects have

Fig. 3.4. Oceanix City, Bjarke Ingels Group (BIG) and Partners, 2019. *Source:* Image courtesy of Oceanix/Big. Oceanix. (n.d.). *Oceanix: Humanitty's Next Frontier*. https://oceanixcity.com/.

also devised floating projects at smaller scales, scales that are not necessarily imbued with the intersection of popular science and geopolitical fantasies. These proposals adapt terrestrial housing typologies to the seas by invoking streamlined and moderne forms, and the white-skinned aesthetics of yacht design over aluminum and steel substructures. For example, in 2015, the architect Giancarlo Zema designed a floating solar-powered home, the WaterNest 100, intended to be made from recycled materials (New Atlas, 2015). Zema's previous projects in a similar vein include Trilobis 65 (2004), a hydrogen fueled hybrid yacht-home, The Neptus 60 Cliff Habitat (2004), a semi-submerged cliff dwelling and the Jelly Fish 45 (2004), a floating apartment block, this was intended to be manufactured by the underwater vehicle specialists UVI (Hanlon, 2004; New Atlas, 2015). Based in Rome, the Giancarlo Zema Design Group, since the early 2000s, has produced a number of "Semi-submerged architectural structures, floating habitats, yacht and interior design" (Giancarlo Zema Design Group, n.d). Zema studied architecture at La Sapienza University under the naval architect Giovanni Zuccon. But oceanic Futurism, if we can call it that, never seems to cease. For example, a more recent floating project by Zema is for EHang, a developer of autonomous electric air vehicles for passengers. This project was designed in 2021 and is a floating vertiport for EHang electric zone taxis in Abu Dhabi (Giancarlo Zema Design Group, n.d).

Another firm that is focused on the niche market of floating architecture is the Netherlands based WaterStudio. The founder Koen Olthuis is a graduate of the Delft University of Technology where he studied both Architecture and Interior Design. Waterstudio's polemic is focused on the intersection of global urbanization and climate change. The work of the studio is focused on the littoral shores, coastlines and cities, as well as the oceans. No doubt the publication of the IPCC AR6 Working Report II addressing coastlines will be referenced in the climate imaginaries of these architects (IPCC, 2022). A number of their projects include The 2019 Sea Tree project, a seemingly biophilic project which is based on the technologies used in the floating storage towers of oil companies. In the proposal, these oil structures are described by WaterStudio as a "'City App'; a floating product that can be added to a city similar to adding an app on your smart phone" (Waterstudio, 2019). The project can be located across a range of aquatic environments including seas, rivers, and harbors. The project includes a series of decks that encourages local ecologies of plants and birds to inhabit above the waterline. Below the waterline, the scheme provides a habitat for sea creatures and when possible artificial coral reefs. The economic rationale for the project is: "The oil company still holds ownership and the city provides a location" (Waterstudio, 2019). Not all of WaterStudio's projects are driven by altruism. For example, the Maldives Floating city, inspired by the morphology of brain coral, is a project with thousands of dwellings available for sale near Male the Capital of the Maldives. As well as deriving its morphology from brain coral, the project is "inspired by traditional Maldivian sea-faring culture" and the "homes will eventually be joined by hotels, restaurants, stylish boutiques and a world-class marina" (Maldives Floating City, 2021).

It would be incorrect, and perhaps all too easy, to suggest that the above salad of signifiers – biomorphism, biophilia, Malthusian ideals, climatic neoliberalism and future-perfect technologies – lie outside of architecture's prevailing discourse. The registers of popular science and marine kitsch, designed in the form of floating monsters, have also been produced by architects firmly entrenched in the prevailing canon of architecture. For example, Oceanix City is a proposal designed by the architecture the Bjarke Ingels Group or BIG. The proposal was championed by the former French tourism minister and the Seasteading Institute; the Seasteading Institute seeks to promote floating "autonomous city-states" (Wang, 2019). Oceanix City employs many of the aesthetic and technological principles evident in the examples above; For instance, systems of modular growth, eco-friendly technologies, and arguments about the advantages of floating cities over land-based urbanism. Oceanix was presented at a 2019 UN-Habitat convened roundtable on floating cities. The roundtable discussed if floating cities could be a "viable solution to urban challenges such as climate change and lack of affordable housing." The meeting was co-convened by Oceanix the company, MIT's Center for Ocean Engineering, and the USA-based Explorers Club (UN-Habitat, 2019). The roundtable aimed to discuss:

> ...the proposed structure and design of floating cities, how they would be situated close to cities and could be used to house those fleeing from rising sea levels and other threats from natural or climate-related disasters in their home areas and overcome housing shortages.
>
> (UN-Habitat, 2019)

Floating cities now appears to be a specialist product line, in BIG's portfolio of projects, with a focus not dissimilar to Waterstudio, on both coastlines and the oceans. However, BIG's projects are much larger – bigger, if we are to state the obvious pun – theatrical pieces of engineering infrastructure as evident in their project in Penang Malaysia PSI Biodiversity City (BIG, 2020). As will be discussed later, the project scale of BIG's ambitions are not limited to the oceans alone (Fig. 3.5).

In contrast to the above, and given the way the above projects are promoted through various media, it is worth considering another frame of reference. The plight of Tuvalu, a micro-nation with a small population, raises issues about adaptation to rising seas, but also issues of how Tuvalu is imagined, and imaged, within the global media system. For example, Carol Farbotko (2005) examined Tuvalu and "the discursive negotiation of the identity of geographically distant islands and island peoples in the Australian news media" (2005, p. 279). They explore and outline the designation of an environmental refugee, noting that representations of "Tuvalu and climate change are routinely produced in non-island places" and how these representations constitute Tuvalu's global identities as a "problematic category" in global climate discourse (2005, pp. 279–280). Concluding that in the Australian media Tuvaluan's are depicted as "tragic victims" through tropes that highlight an "island/mainland alterity" with

Fig. 3.5. Lilypad, Floating Ecopolis for Climate Refugees, Vincent Callebaut Architectures, Oceans, 2008–2017. *Source:* Vincent Callebaut Architectures. (n.da). *LILYPAD: Floating ecopolis for climate refugees, Oceans 2008–2017, World.* https://vincent.callebaut.org/object/080523_lilypad/lilypad/projects.

the effect of limiting and perhaps foreclosing adaptation discourse and operating to "to silence alternative identities that emphasize more empowering qualities of resilience and resourcefulness" (Farbotko, 2005, pp. 288–289). Moreover, this work suggests that adaptive spatial practice and their attendant theories should not be solely centered on ideas of the land but also include the cultural geographies of the sea; this latter element is evident to some extent in Roguerie's Thallasopolis I.

Small countries, like Tuvalu, face many traps in the UNFCC's negotiation and multi-lateral framework. For example, Aykut (2017) describes Tuvalu's interactions with Saudi Arabia in a UNFCCC forum. This asymmetry is exacerbated by the fact that "politics appeared in highly coded forms, disguised as procedural issues, and expressed in the polished language of UN multilateralism" (Aykut, 2017, p. 26). We might contrast this situation with the futurist visions of the architects described above, who appear to have no interest in the cultural geographies of small island nation-states.

What might now be said about the above floating monsters and the climate imaginaries of architects? Japanese film-makers also specialized in using the sea as a kind of immersive fluid in which monsters are gestated. Some of the plots and special effects – or lack of – in these films have become cult classics in their own right. For example *Gamera* versus *Zigra* (1971), in which a giant mutant turtle attempts to prevent Zigra the Deep Sea monster from taking over the Earth (Yuasa, 1971). Similarly, since 2000, like giant mutant turtles, the ecotopian proposals of architects can be seen as technologies of oceanic urbanization and colonization linked to a new ecotopian imaginary. As the above projects establish, these schemes are products that fill a particular niche, as each scheme fashions for itself a brand linked to globalization, neoliberal economics, tourism markets, and the sustainable cities industry. These products accompany a climate-orientated mysticism, ignorant of the sea's ecologies, fueled by a range of identifiable and recurring concepts and compositional techniques. These floating schemes devalue notions of sustainability and eschew a deeper engagement with ecological science. These are architectural monsters that have been ecologically augmented with biophilic gestures and biomimetic computer coding. Like the deep-sea monsters of cinema, these floating monsters take on the familiar tropes and figures seen in popular culture. Consequently, the line between climate science and popular science is blurred in every augmented design gesture.

And in amongst all of all this – notwithstanding Rougerie's attention to the Banda Sea's topography around the Kai islands – there are no storms, no killer whales, no Moby Dick's, no harpoons, no scrimshaw; no Melville or Conrad; there are no sea winds or seasons in the oceans these cities float in. This is a magic that has extracted shapes from the Earth's sea creatures, then multiplied these shapes to create floating monsters; if this magic can do this, we might also ask if it is also the kind of magic that can save living corals from destruction?

Monster Laboratories

Biophysical monsters, shaped by generic tropes of biology, and the biological, also lurk in speculative terrestrial proposals. Ecologic studio is a London-based design studio and its founders, Claudio Pasquero and Marco Poletto, have championed the case of algae as a construction material to alleviate climate change. Since its inception in 2005, Ecologic studio's work has been widely disseminated and published (FRAC Centre, n.db). Notably, its work was exhibited at the Venice Biennale in 2008, 2010, and 2017. The studio has also set up two research labs: the Urban Morphogenesis Lab at the Bartlett University College London, and the other is the Synthetic Landscapes Lab at the University of Innsbruck (Synthetic Landscape, n.d; Urban Morphogenesis Lab, n.d). As a factory of bio-polemics and without any real commissions to speak off, apart from exhibitions, the studio appears to have employed an impressive list of interns from the elite architectural schools such as the AA and UCL. The studio's polemics on the biological sciences and innovations in algae construction have perhaps led to its attraction as a destination for the world's best architecture schools. What follows is a brief account of three Ecologic projects (2005, 2019, 2021), one YouTube presentation (2020) and associated polemics (Fig. 3.6).

Ecologic Projects and Polemics

In the studio's early work, a series of vase-shaped models, Stemcloud v2.0, were exhibited at the Seville Seville Art and Architecture Biennale. The next iteration of this project was Stemcloud v3.0 exhibited at the 2010 Venice Biennale (Ecologic, n.da.; Pohl, 2009). The Seville project was intended to be interactive with the public and make oxygen, and at Seville, and was described to allow:

> ...the process to be visually and materially exposed and interfere with the microclimate of the gallery; the public will feed the colonies present in the river water with nutrients, light and CO2 and as a result oxygenate the gallery space.
>
> (Pohl, 2009)

Fig. 3.6. Figure *STEMv3.0 the Lagoon Experiment*, 2008 Venice Architecture Biennale. *Source:* Author.

In its Venice incarnation Stemcloud v3.0 the studio identifies the problematic of nature conservation which is:

> ...progressively losing its value and needs to be replaced with more adaptive mechanisms of management and direct evaluation of the effects of human transformation of natural ecosystems: we call this mechanism, ecoMachines.
>
> (Bio Tech City, 2014)

As it was exhibited at Venice, the accompanying models are ambiguous in scale: are the towers designed to be housing? Or are they micro-scaled biological machines? They are apparently intended to be small scale models of self-sustaining food factories with water catchments and photosynthesizing tendrils; ecotopias designed to ensure humanities survival in an overheating world. Algae – or as the studio sometimes designates it as cyanobacteria – as a construction material of the future has been a focus of the studio's research and further iterations of these so called ecoMachines – bio-lights, bio furniture, and façade systems (Figs. 3.7 and 3.8).

Ecologic claims to work in the intersection between "biology computation and design" perceiving design as a "morphogenetic process" (Ecologic, n.db). For Ecologic, the aim is to shift toward "a non-anthropocentric view of design" (Ecologic, n.db). Another algae-inspired project developed by the studio with broad ecological ambitions was HORTUS. XL Astaxanthin.g. This project was exhibited at the Center Pompidou in 2019 and claimed to be a "large-scale, high-resolution 3D printed bio-sculpture receptive to both human and nonhuman life." The aim of the project was to give design itself "a new role in redefining ecology" (Ecologic, n.db). In this project, Ecologic seeks to "interrelate between human and machine, human and bacteria and human and algae" and it is with a machinic algorithm through which new relations between biology humans, nonhumans are reconciled, simulated, and created (Ecologic, n.db). This process begins with the making "of a substratum inspired by collective coral morphogenesis" that:

> With a digital algorithm we can simulate the growth of a. This is physically deposited by 3D printing machines in layers of 400 microns, supported by triangular cells of 46 mm and divided in hexagonal blocks of 18.5 cm.
>
> (Ecologic, n.db)

And then, having made the coral-like substratum:

> Photosynthetic cyanobacteria are inoculated on a biogel medium into the individual triangular cells, or bio-pixel, forming the units of biological intelligence of the system.
>
> (Ecologic, n.db)

Fig. 3.7. ecoLogicStudio, HORTUS XL Astaxanthin.g, Center Pompidou, Paris, 2019. *Source:* Architectural Design. Pasquero, C., & Poletto, M. (2019). Beauty as Ecological Intelligence: Bio-digital Aesthetics as a Value System of Post-Anthropocene Architecture. Architectural Design, 89(5), 58–65.

Images of the project, in its different incarnations, indicate a giant plastic syringe with a cake-like piping nozzle inoculating the 3D printed structure with what is described as "living Chlorella cultures on jellified medium" (Ecologic, n.db). In using cyanobacteria in this project, Ecologic claim that the "metabolisms" of the algae are "powered by photosynthesis" which in turn "convert radiation into actual oxygen and biomass." In this project, an algorithm determines in response to "increased incoming radiation," – in other words sunlight – how much of the cyanobacteria are placed into each triangular cell or 3D-printed "bio-pixel" (Ecologic, n.db). Ecologic celebrates the cyanobacteria through these methods of computational design, algae horticulture, and simulation: "Among

Fig. 3.8. ecoLogicStudio, photo.Synt.Etica, Dublin, 2018. Each module functions as a photobioreactor, a digitally designed and custom-made bioplastic container that utilises daylight to feed the living micro-algal cultures. *Source:* Architectural Design. Pasquero, C., & Poletto, M. (2019). Beauty as Ecological Intelligence: Bio-digital Aesthetics as a Value System of Post-Anthropocene Architecture. Architectural Design, 89(5), 58–65.

the oldest organisms on Earth, cyanobacteria's unique biological intelligence is therefore gathered as part of a new form of bio-digital architecture" (Ecologic, n.db).

Not surprisingly algae is also the focus of another Ecologic project presented at the 2021 Architecture Biennale in Venice. The project is the culmination of the studio's "bio-digital" research since its inception. This project is called BIT. BIO. BOT and was exhibited at the "As New Households" space of the Corderie dell'Arsenale. This project pursues the studio's algae obsession at a domestic scale. For this exhibit, the studio constructed a 1:1 scale installation which they describe as an "immersive experiment in the domestic cultivation of the urban microbiome." As with their many other projects the experimental space is designed to test – whatever that might mean – "the co-existence between human and nonhuman organisms in the post-pandemic Urbansphere" (Ecologic, n.dc). As it stood in the Corderie dell'Arsenale, the project is constructed to showcase three different "fluidly interconnected systems that embody the fundamental architectural environments of a future dwelling."

The three systems or elements of the project are as follows: First, "living curtains filled with algae." Second, "large glass tubes" and finally a convivial 2 m by 2 m table that "hosts" 36 pieces of "crystal glassware" (Ecologic, n.dc). The Living cladding can be described as a series of welded vertical curtains 1 m wide by 3 m high. The pattern on these has been designed to reflect the brick walls of the Arsenale itself. Each curtain contains seven liters of algae and is based on Ecologic's own photo; Synthetica façade system (photo.Synthetica, n.d). The second element of the exhibit is the Vertical Garden: "The Vertical Garden is able to absorb CO2 at a rate equal to three large mature trees, providing a clear path to carbon neutrality in architecture" (Ecologic, n.dc). This Vertical Garden is a series of large test tubes constructed from lab-grade borosilicate glass that hang from a stainless-steel frame, and it is within these glass vessels that algae are grown. In this growing medium, the algae can be harvested to yield 100 grams of biomass: "the daily recommended protein intake of a family of four" (Ecologic, n.dc). This allows visitors to the Biennale to experience: "a multiplicity of opportunities to observe, transform and taste microalgae as part of a new culinary landscape" (Ecologic, n.dc).

This glassware is designed by Ecologic and made through 3D printing in conjunction with Swarovsksi. These glass receptacles are made from fused glass layers and are: "designed to encourage experimentation in the consumption of the freshly harvested Chlorella and Spirulina cells, both of which are among the most nutritious organisms on Earth" (Ecologic, n.dc). In creating this new culinary landscape, we can apprehend a sacralization of algae within a secular domestic setting. The performative and active design of the algae in this little theater can be likened to transubstantiation; water into wine; algae into the garden facade; algae into food; algae as the ultimate consumer of CO_2. All these symbols of a new kind of sanctity are reminiscent of the symbolic evocations of the transubstantiation of flesh and blood in the golden chalices monstrances, ciboria, and sacramentaries of Josef Plecnik.

Ecologic's critique of the modernist city, and indeed what the studio calls "the current green paradigm" appears to avoid engaging with issues of power and socio-natures. Ecologic's critique of the city of the twentieth century is equated with the usual suspects of modernist master planning: the evils of functional zoning resulting in segregated zones of production, leisure, and housing. Modernist sanitation is also part of this critique, and this leads to Ecologic identifying the split in modern cities between functional extraction and functional waste. Ecologic name this as the "linear metabolism of the modern city" (Melbourne School of Design, 2020, 29:08). For Ecologic this is a city acting and produced through "a machine-like idea of nature dominated by the idea of equilibrium" (Melbourne School of Design, 2020, 7:01). Looking at the studio's corpus, we witness a number of common devices and figures already present, and easily recognisable, in the climate discourse of architects. This is a polemic with an associated aesthetics based on a misguided faith in the signifiers of conventional science: the laboratory, the authority of cybernetics, the mapping of complex systems, and seemingly scientific naming conventions. As with any propaganda, this integration of tropes heralds, for Ecologic at least, a new era which changes the very basis of human existence: by bringing together in their words "the dictates of human rationality with the effects of proximity to bio-artificial intelligence" (Ecologic, n.db). Everything that ever existed will be part of this new world including:

> …electronic devices, digital algorithms, many forms of plant and animal changes, bacteria, but, of course, also, viruses, let them become what we call bio-citizens, actively involved in the formation and transformation of the human landscape.
>
> (Biennale Architettura, 2021)

For both humans and the cyanobacteria who will become bio-citizens, there will be no need for governance or politics in this new world. Ecologic's bio-polemic is intent on creating a system of biological production. Algae is simultaneously plant, construction material, food product, and absorber of CO2 – all combined into one system. Ecologic's project is to hope for a world biology, an ecological world system with algae at its center. At every point Ecologic's gestures, polemics and projects seek to expand in scale and configure a new system based on the biology of micro-organisms. In their own words they ask:

> So, how will we live together? Let's imagine an expanded notion of the We to include alongside humans the plethora of complex synthetic organisms currently inhabiting what we call the urbansphere.
>
> (Biennale Architettura, 2021)

The broader Ecologic project is one that is ultimately about beauty and climate imaginaries rather than the pursuit of carbon mitigation. Ecologic hopes to expand aesthetic experience through new modes of pattern recognition and rationality, "thus expanding the practice's repertoire of aesthetic qualities"(p. 139). But the terms of this aesthetic project is redefined: "Aesthetic here is intended as a meta-language, enabling a more sophisticated level of communication with the nonhuman." For Ecologic architecture takes on a new role, a role that is "no longer a case of architecture being inspired by other disciplines, such as biology and computer science, and striving to become biomimetic of biophilic" (Pasquero & Poletto, 2020, p. 139). In other words, a role that is no longer based on simple mimesis but one that hopes to integrate and develop a new language between architecture, biology, the algorithms, and, of course, the algae.

What then can we say about Ecologic's theatrical laboratories? In these theaters of polemical production – despite the rhetoric of biophysical integration – aesthetics still reigns supreme and is worshipped through the algae. Of course, if one familiar figure looms large and haunts the work of Ecologic, it is Buckminster Fuller: planetary engineer, global autocrat, and salesman. But there is also the all too familiar notes of Darcy Thompson's diatoms and Frei Otto's soap bubbles of the early 1960s. But there is also another darker architectural precedent we might propose. Leon Krier and his relationship to Speer's classicism is not so far removed from Ecologic's ever expanding urban sphere and the orders of the algae: A systemic architectural language of cyanobacteria, embodying a new take on the Vitruvian firmness commodity and delight (Krier, 2013). In other words a new classicism of algae. A biodigital and parametric classicism and corresponding aesthetics, with its own botanical orders, that has no need of cultural codes or designations of power. This is because in Ecologic's project, living systems are extracted and distilled as algae – simulating far distant future technologies – apparitions that suggest a future bio-industrialization of the Earth's cities. These cities of biological production will operate as open systems. Ecologic, rather than bringing an errant nature back into equilibrium, will bring an errant nature back into a new controlled open system of production, a new city where both capitalist production and consumption will continue its violence without biological limits.

Conclusion

As a response to global warming, ecotopian cities exemplify the desire to forever extend the urban sphere. The digital cities of the computer, generated by scripts and algorithms, have empowered these imagined and phantom cities of a future Anthropocene that will never exist. As a result, there has been a resurgence in the architectural imagination of designs for these ecotopian cities; cities exemplified by imaginary lines of flight into the deserts, the seas, and even outer space. In these cities, there is a nostalgia for Ernest Haeckel's Radiolaria; as radiolaria structures have come to inhabit the immersive oceans and outer realms of digital architecture. In these biophysical and climate imaginaries, existing cities, and developments within existing cities, have been populated and smeared with green colored organic matter. In all of these imagined cities we find a curious mix of market derived technologies, wedded to biomorphic forms and yet the spectre of a climate force-fed with greenhouse gases remains.

Reproduced on the Instagram account of Ryan Scavincky, an architecture critic with around 10,000 followers, is a post that depicts a contrast between the worldview of scientists and the worldview of architects. It states: "Scientists: 75% of the greenhouse gases in the atmosphere at this very moment will still be there in half a millennium" (Scavnicky, 2019). And then, as the Instagram post posits it, Architects respond by producing an image of a luxuriant and green saturated floating city on the sea. The city is on the edge of what seems like a tropical paradise and contains a marina filled with boats, smaller breakwaters with swimming pools, and conch-like housing or hotel blocks. The image depicts a utopian world caught between land and the oceans where both appear to co-mingle. This Instagram meme points to the way that architects have responded in the recent past to the unfolding climate emergency. These types of responses – a future-perfect Anthropocene – point to a corrosive streak of poetic imagination within architectural discourse that simultaneously produces and generates an imaginary narrative that bears little relationship to science. With the advent of the very real prospect of human extinction because of catastrophic climate change, the poetic response, and its associated image politics, must be urgently questioned.

As if to call a stop to all of the above, Antoine Picon reinforces the point that the climate emergency has not been a part of the narrative of those architects who might be described as part of a digital avant-garde (Picon, 2020). Writing in *Architectural Design* in 2020, Picon writes:

> Strangely, given its crucial importance in our anxious and endangered contemporary world, climate change has not been very present in this story, as if issues of thermal exchanges, grey energy and rising waters were peripheral to the revolution announced by digital avant-gardes, including the present-day advocates of artificial intelligence.
>
> (Picon, 2020, p. 120)

Invoking Buckminster Fuller, Picon returns to the cyborg, the plumbed up human, as a way out of the dilemma of our times, arguing that a cyborg world can be seen as a:

> …complex networked ecologies that blur the distinction between the self and its environment and make obsolete the notion of a clear-cut boundary between subjective interiority and a purely objective exterior.
>
> (Picon, 2020, p. 122)

Picon's solution, if we can call it that, to bridge the gap between the parametricism of digital architects is to call for a return to materiality. In doing so, Picon makes a call to designers to consider the underlying and essential "relationship between humans and the material world" (Picon, 2020, p. 124). For Picon bridging the gap across the material and the digital will create and connect to new "forms of subjectivity." Picon (2020) cites the work of Achim Menges as an exponent of experiments that have successfully blurred matter and digital computation. He sees this turn to the material, in a reassessment of ornament, ornament being a reification of a physical work resulting in an architectural realm of the senses, a "sensorium," where architects and in the fabrication labs of architecture schools find a common denominator between environmental and digitally orientated designers; Picon suggests that both are "exploring the same world' and that materials a now a new frontier of design (p. 124). In attempting to bridge the gap, the scales of global-local divide Picon, in order to go beyond "the seduction of digital virtuosity and environmental moralism." Picon resorts to quoting Heidegger's Building Dwelling Thinking – perhaps forgetting that Heidegger was a Nazi (p. 125). In any case, the techno-nature proposed in this approach is still overly anthropocentric in its outlook and only promotes what might be called a path toward a good Anthropocene in an imagined equilibrium.

All this illustrates the historical outlines of the biophysical in contemporary architecture. Certainly, a city inhabited by hordes of human cyborgs and Robbie the Robots, and their infrastructural variants, may only ever create a fragmented biophysical city. A city always reaching for a biophysical unity and little else. These cities might only ever be cities whose neo-neo-futurist guises circulate in the delirium of architectural discourse. And if any reminder is needed about the dangers or attributes of neo-futurism, it is worth noting what Rowe and Koetter (1983) were to remark about Futurism: "the celebration of force is among its more important sustaining sentiments" seeing a parallel between Nietzsche's "the free man is a warrior" and Futurism's "intrinsic atavism"; replete with "phallic fantasies" and proto-Fascism (Rowe & Koetter, 1983, p. 30).

References

Aykut, S. C. (2017). Governing through verbs: The practice of negotiating and the making of a new mode of governance. In S. C. Aykut, J. Foyer, & E. Morena (Eds.), *Globalising the climate* (pp. 18–38). New York, NY: Routledge.

Batty, M., & Hudson-Smith, A. (2005). Urban simulacra: London. *Architectural Design*, *75*(6), 42–47.

Beekmans, J. (2012, June 21). The wave garden. *Archis*. Retrieved from http://archis.org/volume/the-wave-garden

Biennale Architettura. (2021, February 3). Sneek peek: EcoLogic studio. [video]. *YouTube*. Retrieved from https://www.youtube.com/watch?v=GS4pcdIVajc&t=4s

BIG. (2020). PSI biodiversity city Penang. Retrieved from https://big.dk/#projects-psi

Bio Tech City. (2014, December 14). ArtEcologies. [blogpost]. Retrieved from https://biotechcity.tumblr.com/post/105168913984/artecologies

CCA. (n.d). Water flux and scrambled flat (2002–2010). Retrieved from https://www.cca.qc.ca/en/archives/464863/rsien-project-records/500302/water-flux-and-scrambled-flat

Dahinden, J. (1972). *Urban structures for the future*. London: Pall Mall Press.

Ecologic. (n.da). Stemcloud. Retrieved from https://www.ecologicstudio.com/projects/stemcloud-seville-art-and-architecture-biennale-2008

Ecologic. (n.db). H-O-R-T-U-S-XL-ASTAXANTHIN-G. Retrieved from https://www.ecologicstudio.com/projects/h-o-r-t-u-s-xl-astaxanthin-g

Ecologic. (n.dc). BIT.BIO.BOT. Retrieved from https://www.ecologicstudio.com/projects/bitbiobot

Fox, S. (2010, March 12). Concept waterscraper brings *monumental architecture into the open sea. Popular Science*. Retrieved from https://www.popsci.com/technology/article/2010-03/concept-water-scraper-brings-monumental-architecture-open-sea

Farbotko, C. (2005). Tuvalu and climate change: Constructions of environmental displacement in the Sydney Morning Herald. *Geografiska Annaler: Series B, Human Geography*, *87*(4), 279–293.

FRAC Centre. (n.da). Megastructures. Retrieved from https://www.frac-centre.fr/ressources/parcours-thematiques/megastructures/megastructures-619.html

FRAC Centre. (n.db). ecoLogicStudio (Claudia Pasquero, Marco Poletto). Retrieved from https://www.frac-centre.fr/_en/art-and-architecture-collection/ecologicstudio-316.html?authID=358

Frichot, H. (2020). *Dirty theory: Troubling architecture*. Spurbuchverlag: AADRL Spurbuchverlag.

Gian Carlo Design Group. (n.d). Gian Carlo Design Group. Retrieved from https://www.giancarlozema.com

Gruskoff, M. (Producer) & Trumball, D. (Director). (1971). *Silent running*. Universal City, CA: Universal Pictures.

Hanlon, M. (2004, June 4). Floating apartment offers *deep blue vista. New Atlas*. Retrieved from https://newatlas.com/go/1887/?itm_source=newatlas&itm_medium=article-body

Haraway, D. (1991). *Simians, cyborgs and women: The reinvention of nature*. New York, NY: Routledge.

Hunter, P., & Harrison, A. (n.d.). Preliminary inventory of the Yona Friedman papers, 1956–2006. *Online Archive of California*. Retrieved from https://oac.cdlib.org/findaid/ark:/13030/c8p84hgg/entire_text/

Institute for Computational Design and Construction (ICD). (2019). Urbach tower -ICD research buildings/prototypes Remstal Gartenschau 2019. Retrieved from https://www.icd.uni-stuttgart.de/projects/remstal-gartenschau-2019-urbach-turm/

IPCC. (2022). Climate change 2022: Impacts, adaptation, and vulnerability. In H.-O. Pörtner, D. C. Roberts, M. Tignor, E. S. Poloczanska, K. Mintenbeck, A. Alegría, … B. Rama (Eds.), *Contribution of working group II to the sixth assessment Report of the intergovernmental panel on climate change*. Cambridge: Cambridge University Press.

Jencks, C. (1971). *Architecture 2000: Predictions and methods*. New York, NY: Praeger.

Kaji-O'Grady, S., & Raisbeck, P. (2005). Prototype cities in the sea. *The Journal of Architecture*, *10*(4), 443–446.

Knippers, J., Nickel, K. G., & Speck, T. (2016). *Biomimetic research for architecture and building construction* (pp.1–5). Cham: Springer International Publishing.

Krier, L. (2013). *Albert speer: Architecture 1932–1942*. New York, NY: Monacelli Press.

Maldives Floating City. (2021). Maldives Floating City. Retrieved from https://maldivesfloatingcity.com

McGaw, J. (2018). Dark matter. *Architectural Theory Review*, *22*(1), 120–139.

Meinhold, B. (2010a, February 9). Futuristic floating city is an Ecotopia. *Inhabitant.* Retrieved from https://inhabitat.com/futuristic-floating-city-is-an-ecotopia-at-sea

Meinhold, B. (2010b, August 2). The Gyre: A floating upside-down eco skyscraper. *Inhabitant*. Retrieved from https://inhabitat.com/the-gyre-a-floating-upside-down-eco-skyscraper

Melbourne School of Design. (2020, February 14). BIT. BIO.BOT – Dr Marco Poletto [video]. *YouTube*. Retrieved from https://www.youtube.com/watch?v=KnOMlKdy3yY&t=5s

Migayrou, F., & Mennan, Z. (2003). *Architectures non standard.* Paris: Editions du/Centre Pompidou.

New Atlas. (2015, March 13). Giancarlo Zema makes a splash with his eco-friendly solar-powered floating home. Retrieved from https://newatlas.com/giancarlo-zema-ecflolife-waternest-100/36534

new-territories. (n.da). new-territories/[Laboratory of digital disobedience]/R&Sie(n)/[eIf/bʌt/c]/M4 (mindmachinemakingmyths). Retrieved from https://www.new-territories.com/roche%20cv.htm

new-territoires. (n.db). Dustyrelief/wireframe. Retrieved from https://www.new-territories.com/unplug.htm

Pasquero, C., & Poletto, M. (2020). Bio-digital aesthetics as value system of post-Anthropocene architecture. *International Journal of Architectural Computing*, *18*(2), 120–140.

photo.Synthetica. (n.d). photo.Synthetica. *ecologicalStudio*. Retrieved from https://www.photosynthetica.co.uk/

Picon, A. (2020). Beyond digital avant-gardes: The materiality of architecture and its impact. *Architectural Design*, *90*(5), 118–125.

Pohl, E. B. (2009, July 23). STEMcloud v2.0/ecoLogicStudio. *ArchDaily*. Retrieved from https://www.archdaily.com/29718/stemcloud-v20-ecologicstudio

PREVI/Lima Low Cost Housing Project. (1970). *Architectural Design*, *40*(4), 187–205.

Pruned. (2005, June 18). Wave garden by Yusuke Obuchi. [blogpost]. Retrieved from https://pruned.blogspot.com/2005/06/wave-garden-by-yusuke-obuchi.html

Ragon, M. (1986). *Histoire mondiale de l'architecture et de l'urbanisme modernes.* Tournai: Casterman.

Retalk Asia. (n.d., September, 22). Self-sufficient eco-tourist resort designed to revitalise the health of the Philippine Sea. Retrieved from https://www.retalkasia.com/property/nautilus-eco-resort-palawan/self-sufficient-eco-tourist-resort-designed-revitalise-health retrieved 211021

Roche, F. (2012). Reclaim Resi [lience] stance//...... R^2. *Log*, *25*, 1–5.

Roche, F., & Lacadée, C. (2016). MMYST vs. concrete [i] land. *Log*, *36*, 75–86.

Rougerie, J. (1972a). Thalassopolis. *L'architecture D'aujourd'hui*, *164*, 80–83.

Rougerie, J. (1972b). Thalassopolis. *Urbanisme*, *41*(133), 34–45.

Rougerie, J. (1974). Thalassopolis. *L'architecture D'aujourd'hui*, *175*, 70.

Rougerie, J. (1977). The architecture of the sea: Integration of man in the aquatic environment. *Industrialisation Forum*, *8*(4–5), 66.

Rougerie, J., & Vignes, E. (1978). *Habiter la mer*. Paris: Editions maritime et d'Outre-Mer.

Rowe, C., & Koetter, F. (1983). *Collage city*. Cambridge, MA: MIT Press.

Scavnicky, R. [@Sssscavvvv]. (2019, October 4). Scientists: 75% of the greenhouse gases in the atmosphere at this very moment will still be there in half a millennium. *Instagram*. Retrieved from https://www.instagram.com/p/B3LhNuDJ2hV

Shizimu. (n.d). The environmental Island, green float. Retrieved from https://www.shimz.co.jp/en/topics/dream/content03

Smithson, A. (1974). How to recognise and read mat-building. *Architectural Design, 9*(September), 573–590.

[SA] Squared. (n.d). Links. Retrieved from https://www.sarly.net/links

Synthetic Landscape. (n.d). Synthetic landscape. Institute of urban design, university of Innsbruck, prof. Claudia Pasquero. [video]. *YouTube*. Retrieved from https://www.youtube.com/channel/UCglbiyUL1f_jfBuyE9yU7vQ/videos

Thompson, D. W., & Thompson, D. A. W. (1942). *On growth and form* (Vol. 2, pp. 470). Cambridge: Cambridge University Press.

UN-Habitat. (2019, April 3). Roundtable on floating citie*s at UNHQ calls for innovation to benefit all.* Retrieved from https://unhabitat.org/roundtable-on-floating-cities-at-unhq-calls-for-innovation-to-benefit-all

Urban Morphogenesis Lab. (n.d). About. Retrieved from https://urbanmorphogenesislab.com/about

Venus Project. (n.d). Cities in the sea. Retrieved from www.thevenusproject.com/resource-based-economy/environment/cities-in-the-sea

Vincent Callebaut Architectures. (2015). AEQUOREA: Oceanscraper printed in 3D from the seventh continent's garbage, the 5 ocean gyres. Rio De Janeiro 2015. Retrieved from https://vincent.callebaut.org/object/151223_aequorea/aequorea/projects

Vincent Callebaut Architectures. (n.da). LILYPAD: Floating ecopolis for climate refugees, oceans 2008–2017, world. Retrieved from https://vincent.callebaut.org/object/080523_lilypad/lilypad/projects

Vincent Callebaut Architectures. (n.db). Vincent Callebaut architecture. Paris, France (CV). Retrieved from https://vincent.callebaut.org/cv

Vincent Callebaut Architectures. (n.dc). About us. Retrieved from https://vincent.callebaut.org

Wang, B. T. (2019, June 3). Floating cities: The future or a washed-up idea? *The Conversation*. Retrieved from https://theconversation.com/floating-cities-the-future-or-a-washed-up-idea-116511

Waterstudio. (2019). Sea tree. Retrieved from https://www.waterstudio.nl/projects/sea-tree

Whitham, G. (1990). Science fiction. In D. Robbins (Ed.), *The independent group: Postwar Britain and the aesthetics of plenty*. Cambridge, MA: MIT Press.

wpengine. (2010, November 1). Green float. The botanical city *concept*. Seasteading Institute. [Blogspot]. Retrieved from https://www.seasteading.org/green-float-the-botanical-city-concept

Yoneda, Y. (2011, February 15). Underwater skyscraper is a self-*sufficient CITY at sea. Inhabitat*. Retrieved from https://inhabitat.com/underwater-skyscraper-is-a-self-sufficient-city-at-sea

Yuasa, N. (Director). (1971). Gamera vs. Zigra. *Daiei Film and Dainichi Eihai.*

Chapter 4

Cities in the Global South

Introduction

It is now timely to move from the biophysical deliria of architects to a consideration of the cities of the Global South. These cities are the developing cities, the subaltern cities, the cities that inhabit the invisible continents of the architectural imagination. The cities of the Global South are very different to the future green cities and climate imaginaries of the architects of the Global North. In a 2011 article entitled *Slumdog cities: Rethinking subaltern urbanism*, Ananya Roy was to designate the cities of the Global South as subaltern cities in order to cleave open "subaltern" spaces and retrieve these cities from a mire "of apocalyptic and dystopian narratives of the slum" (Roy, 2011, p. 223). An informal settlement is usefully defined as a place where there is no permanent tenure, housing, adequate living spaces, or access to water, sanitation, and drainage systems. UN-Habitat III (2015) more fulsomely defines these settlements in the following terms:

> Informal settlements – are residential areas where 1) inhabitants have no security of tenure vis-à-vis the land or dwellings they inhabit, with modalities ranging from squatting to informal rental housing, 2) the neighbourhoods usually lack, or are cut off from, basic services and city infrastructure and 3) the housing may not comply with current planning and building regulations and is often situated in geographically and environmentally hazardous areas. (p. 1)

From this definition emerges the UN-Habitat III definition of a slum:

> Slums are the most deprived and excluded form of informal settlements characterised by poverty and large agglomerations of dilapidated housing often located in the most hazardous urban land. In addition to tenure insecurity, slum dwellers lack formal supply of basic infrastructure and services, public space and green areas, and are constantly exposed to eviction, disease and violence.
> (UN-Habitat III, 2015, p. 1)

Architects, Sustainability and the Climate Emergency, 85–115

Published under exclusive licence by Emerald Publishing Limited
doi:10.1108/978-1-80382-291-420221004

For many urban scholars, theorists, and policymakers, a fraught and evolving discourse surrounds the above terminology. Dovey, van Oostrum, Chatterjee, and Shafique (2020) rage at the fact that informal settlements cannot be conflated with the word slum, a term easily employed as a weapon of demolition policies – a point agreed with herein. Informal settlements are better seen as a mode of production: "a verb rather than a noun" (Dovey et al., 2020, p. 1; Dovey, Shafique, van Oostrum, & Chatterjee, 2021). As Fran Tonkiss notes informality is better considered as an idiom – from morphologies to practices, grammar to idiom – of how space is produced (Redento Recio, 2021a, 4:36). These distinctions and terms are important to debate because of the scale of urbanization in the Global South. As asserted in the (UN-Habitat, 2020) Climate Change Vulnerability and Risk Community Assessment guide:

> Urbanisation is one of the global megatrends of our time, unstoppable and irreversible. In 30 years, two-thirds of the world's population will live in urban areas; 90% of this urban growth will take place in less developed regions such as East Asia, South Asia, and Sub-Saharan Africa. These are regions where capacity and resources are most constrained, and development challenges are ever more complex and concentrated.
>
> (UN-Habitat, 2020)

The severity of the climate emergency, in tandem with a changing climate, alongside rapid urbanization in the developing world is unarguable. This suggests that theories of urbanization and climate are better formulated in relation to informal settlements than formulated on the premises of the cities of the Global North. Developing cities and informal settlements are integral to considering how the planet responds to the climate emergency. These cities demand attention before examining the so-called green idioms and modes of spatial production evident in the projects of the architects of the Global North. As discussed in the next section, the UN-Habitat's Sustainable Development Goals (SDG) are a primary lens for considering the cities of the Global South.

The Sustainable Development Goals (SDGs) and the New Urban Agenda (NUA)

Certainly, the 17 SDGs are together a behemoth of objectives that can easily overwhelm because of their range, depth, and complexity as human goals. Nonetheless, the implementation of the SDGs is key to identifying and indexing how architects have engaged with developing cities and climate. Arguably, one virtue of the SDGs is that the top-down structure, governance, and logics of the 17 goals comprise a holistic vision of sustainable development. SDG 11 is explicitly focused on urban issues and recognizes the role of cities in urban development. Each goal is interdependent with other goals and for each goal there are aspirational targets and indicators. Moreover, some goals appear to be paradoxical, necessitating trade-offs, when placed side by side with each other (Valencia et al., 2019, p. 5). Habitat I took place in Vancouver Canada in 1976, and this resulted in the Vancouver Action Plan as well as the UN Centre on Human Settlements. This centre later became UN-Habitat (Stiftel, 2021, p. 3). Given this, to what extent have architects themselves operated under the umbrella of UN-Habitat and engaged with the SDGs – if at all? Have architects, both in theory and practice, been enmeshed in the SDGs and their implementation? Thus, at this point It is also worth listing the SDGs in full below:

Goal 1. End poverty in all its forms everywhere
Goal 2. End hunger, achieve food security and improved nutrition, and promote sustainable agriculture
Goal 3. Ensure healthy lives and promote well-being for all at all ages
Goal 4. Ensure inclusive and equitable quality education and promote lifelong learning opportunities for all
Goal 5. Achieve gender equality and empower all women and girls
Goal 6. Ensure availability and sustainable management of water and sanitation for all
Goal 7. Ensure access to affordable, reliable, sustainable, and modern energy for all
Goal 8. Promote sustained, inclusive, and sustainable economic growth, full and productive employment, and decent work for all
Goal 9. Build resilient infrastructure, promote inclusive and sustainable industrialization and foster innovation
Goal 10. Reduce inequality within and among countries
Goal 11. Make cities and human settlements inclusive, safe, resilient, and sustainable
Goal 12. Ensure sustainable consumption and production patterns
Goal 13. Take urgent action to combat climate change and its impacts
Goal 14. Conserve and sustainably use the oceans, seas, and marine resources for sustainable development
Goal 15. Protect, restore, and promote sustainable use of terrestrial ecosystems, sustainably manage forests, combat desertification, and halt and reverse land degradation and halt biodiversity loss
Goal 16. Promote peaceful and inclusive societies for sustainable development, provide access to justice for all and build effective, accountable and inclusive institutions at all levels
Goal 17. Strengthen the means of implementation and revitalize the Global Partnership for Sustainable Development (UN, 2017).

Interconnected with the SDGs is the New Urban Agenda (NUA). The NUA was instigated at UN-Habitat III at Quito in 2016. The NUA was the result of a process that involved extensive participation by civil society stakeholders (Stiftel, 2021, p. 1). The NUA places cities as the key site of the Earth's sustainable development; hence the role of cities is well recognized as being a central part of the SDGs and NUA research, policy, and implementation. Valencia et al. (2019) rightly identify that the NUA was an important point in UN history because it was the first time that the role of subnational bodies, such as local governments, had been acknowledged in terms of sustainable development (p. 5).

The NUA emphasizes the importance of planning across the sectors of "urban design, transport, housing, economic development, environment, health and basic services" (p. 5). These are related to five scales of planning, including "urban design, transport, housing, economic development, environment, health and basic services" (p. 5). The actors who operate within this strategic and conceptual system are described as "national governments, local authorities, planning professionals and civil society organisations" (p. 5). All these entities act with three primary instruments of "urban rules and regulations, urban planning and design, and municipal finance" (Stiftel, 2021, p. 5).

Sustainability Transitions

From the above, it can be seen that the following goals 3, 6, 7, 8, 9, 11, 12, and 15 are directly related to architects and spatial production, rather than simply its regulation, across the built environment. Much of the work around the SDGs, sustainability transitions and the Global South allows a landscape to be explored that too often lies outside of the climate imaginaries and dream-like biophysical theming of architects. But the point of this assertion is the need to be clear-sighted in the face of climate catastrophes. As Köhler et al. (2019) write in a survey of sustainable transitions research, climate change, biodiversity loss, and natural resource depletion is related to unsustainable progress and that this situation is unable to be lessened "by incremental improvements and technological fixes" – as suggested in following chapters (p. 2).

A consideration of the SDGs potentially allows for a more granular view of architecture and sustainability transitions. In eschewing a singular focus, on particular technical fixes or design theories, for example, a more detailed narrative of how architects work – and do not work – in the cities of the Global South emerges. To reiterate the theories and methodologies of the SDGs landscape and sustainability transitions point to how architects might pursue sustainability beyond a focus on technical fixes, one-line global solutions, or virtue signalling – evident in the legacy of Buckminster Fuller and John Mchale, who huckstered their way around the 1960s (Vidler, 2008).

If sustainability transitions appear to be locked out of architectural theory, then there are also other domains locked out of the sustainability discourse. For example, the predominant architectural narratives of the Global North since the 2000s have ignored cities in Africa. Indeed, there have been calls for the use of urban data, by architects and urbanists, to better understand urbanization (Kvan, 2020). However, very little of this work has examined urbanization processes on the African continent. To counter this situation, Africapolis, as it has been designated by the Sahel and West Africa Club (SWAC), now comprises an extensive database that documents urbanization in Africa (Africapolis, 2021a).

This database is intended to understand the challenges of increasing urbanization, "the fastest urban growth rate in the world" – across the continent (Africapolis, 2021a). Africapolis combines both demographic data with spatial imagery, and as a result a more nuanced view of African cities is seen. By employing multiscaled analytics that does not simply privilege megacities and settlements. Africapolis notably makes visible those smaller cities and settlements that might otherwise be invisible in databases. As noted by Africapolis:

> Africa has 11 urban areas with more than five million inhabitants, which have a combined population of 95 million people. The continent also counts 6,740 small urban agglomerations with fewer than 100 000 inhabitants, which are home to 180 million population.
>
> Africapolis (2021b)

Moreover, the concept that databases – as well as architectural theories – can also make particular settlements both visible and invisible is a necessary point. In what follows we might ask how easily do the projects and spatial practices of the Global South lend themselves to having a data presence – a presence in databases? Research focused on the SDGs – sustainability transitions or transition design – where "sustainability transitions theories are integrated with design theory, education and practice" have been far removed from architectural research and discourse (Gaziulusoy & Erdoğan Öztekin, 2019, p. 1). The general characteristics of transition research raise important theoretical and methodological questions for architects. Firstly, the object of analysis in SDG, sustainability, and transition studies is not global scales and visions, nor top-down concepts such as capitalism, or nature-society binaries, or positing sustainability-orientated fixes in cities (Köhler et al., 2019). For architects, sustainability transition research suggests there are real questions about detailed or generic articulation; should complexity be read in a way that simplifies it or leaves its inherent detail unstructured? Should the focus be on seemingly breakthrough technologies or fixes, or should it focus on "process patterns of decline and destabilization?" (Köhler et al., 2019, p. 19).

Significantly then, the SDGs and NUA link to sustainability narratives in fields as diverse as urban theory, planning, development studies, and innovation systems. For architects this points to new theoretical narratives and agencies related to community participation and collaboration, coproduction with other experts and policy cocreation. These new narratives are desperately needed in architectural discourse. And such narratives might create new creative and spatial practices; shifting sustainability transitions away from subject formations that embody a litany of sins: seemingly scientific solutions, overt technocratic and managerial framings; and at the end of the spectrum, poetic visions that only produce hallucinatory realities. These are sins that can only contribute to ecological catastrophe.

Cities and Architects in the Global South

Of course, there are some individual architects and urbanists who have proposed theories regarding the urbanization of cities in the Global South. For example, Rahul Mehrotra is the founding principle of RMA Architects, and both a graduate from, and a professor of Urban Design and Planning at Harvard Graduate of Design. Moreover, Mehrotra is a trustee of the Urban Design Research Institute (UDRI), founding member of the Indian Institute for Human Settlement (IIHS) and Architecture Foundation India; he is thus an activist, educationist, architect, and an urban planner (RMA Architects, n.d.a). Notably, among many other exhibitions, he was responsible for the State of Architecture 2016 exhibition, an exhibition in Mumbai that surveyed the architectural profession in India from 1947 to 2016. The exhibition asked and addressed the questions "Does architecture matter?" Are architects still relevant in India, and can they contribute in any significant way to a nation-state and a society in extraordinary flux? (RMA Architects, n.d.b). With these credentials, Mehrotra, with nearly 15k Instagram followers, has a global audience.

Mehrotra's own polemic is one that is against architecture. Instead, Urban Design is reinforced as a bridging discipline that counters what they perceive as the spectacle of architectural form. Mehrotra pursues this polemic through the concept of the kinetic city, an approach not dissimilar to Urban Think Tank discussed below. Mehrotra describes this kinetic city as being in "a state of constant flux" (Mehrotra & Vera, 2017, p. 44). This kinetic city is not simply a city constantly moving, nor one that is not "two-dimensional" but instead a "multi-faceted" entity, a three-dimensional entity that is in constant motion. In this theory of the kinetic city, configuring functional programs is a primary concern; an approach that is not simply "the construction of the architectonic body" but instead, in this kinetic city "openness and flexibility prevails over rigidity and flexibility is valued over rigor" (Mehrotra & Vera, 2017, p. 46).

In addressing climate change, Mehrotra argues that the urbanist must plan for both "slow and rapid disasters" as this accounts for varying impacts of climate change over different time scales (Das, 2021). In uncoupling architectural form from functional and temporal processes, sustainability is then predicated on a "city's capacity to deconstruct, disassemble, reconfigure, and reverse previous iterations." For Mehrotra architecture is firmly embedded in the capitalist city, with each building operating as its fundamental component, a basic building block of urbanism, that operationalizes the city as the central theater for flows of capital (Mehrotra & Vera, 2017, p. 44). In Mehrotra's own words, architecture is "the basic unit of urbanism, seems to be obsessed with the idea of the city as the centralising spectacle often driving the inherent impatience of capital" (2017, p. 44). And under this practice of architecture, aligned with neoliberalism, urban design is squeezed out of the equation and "architecture becomes the sole instrument through which cities are imagined and formed" (Mehrotra, 2020, p. 132). Against this, Mehrotra posits the idea of ephemeral or temporary landscapes: "the temporary landscape opens a potential space for questioning the idea of permanence as a univocal solution to various urban conditions" (Mehrotra & Vera, 2017, p. 45).

And the above in turn, creates "heterotopic spaces in which one can usually see the dissolution of social predetermination and a diversity that is not normally achieved in the everyday city" (2017, p. 48). Expanding beyond normative and technocratic visions of the city, he proposes a taxonomy of temporary landscapes. This taxonomy of impermanent and ephemeral landscapes includes refugee camps, military encampments, disaster camps, camps for the extraction of resources and what he terms as ephemeral-extractive spaces. Mehrotra's polemic also refers to and celebrates Burning Man and Glastonbury. More interestingly, this logic of temporality draws on the Italian urban theorist and cofounder of Archizoom Andrea Branzi's "New Athens Charter." (Vera & Sordi, 2015) Mehrotra claims that this charter:

> ...offered critical provocations for the radical rethinking of the urban as a more inclusive and less Eurocentric project.
>
> (Mehrotra & Vera, 2017, p. 46)

Directly quoting Branzi's thought from Vera and Sordi (2015) Mehrotra argues a new city can emerge predicated on:

> ...the construction of a new concept of time in a city that should be thought of as a place for 'cosmic hospitality' in which designers could encourage 'planetary coexistence' between 'man and animals, technology and divinities, living and dead.' (p. 46)

And consequently:

In doing so, a model for the city could be found that is "less anthropocentric and more open to biodiversity, to the sacred, and to human beauty" (p. 46).

Through these theoretical mechanisms, the negation of architecture, the privileging of the temporary, the reconceptualization of time, Mehrotra is able to loosen urban designers from the control of a technocratic and financial driven spectacle. A spectacle whose only intent is to create architectural icons and monuments. This is an approach that seeks to renew the city through a kind of anticolonialism, an escape from "contemporary modalities of conquest" ((Mehrotra & Vera, 2017, p. 54). Thus creating new productive processes and efficiencies that do not directly lead to the disproportionate and extreme accumulation of capital. Through this polemic Mehrotra seeks to ensure that the city becomes its own engine of life. Where, as he enigmatically states it, a new kind of city will come into being "that constructs the conditions for life itself to thrive, for 'action' to retreat to become a secondary human activity" (Mehrotra & Vera, 2017, p. 54).

In this fashion Mehrotra's polemic does not simply drift into a litany of open flexibility and unending growth as it does for the exponents of complex adaptive systems – as discussed in the next chapter. Mehrotra's kinetic city is a celebration of the Global South cities of the world, recognizing new urban conceptions of time, the sacred, the ecological and the nonhuman. Mehrotra's vision of cities, with its homage to Branzi, and hints of SuperStudio and Constant's New Babylon, suggests an urbanity where the temporality of play is unfettered by the iconic certainties of tectonic objects, permanent settlement, and unending growth.

The architect and urbanist Alfredo Brillembourg, founder of Urban Think Tank (UTT), in contrast to the multiscalar analytics of Africapolis and Mehrotra's antiarchitecture, fashions an urban theory of the city from the project and site-specific level up. Founded in Caracas Venezuela, UTT's book *Informal city: Caracas case* was published in 2005 (Brillembourg, Feireiss, & Klumpner, 2005). Brillembourg describes UTT's objectives as pursuing design "not as a hermetic self-referential pursuit" but to advance architecture's social agency through "undertaking site cultural specific works" (Redento Recio, 2021, 3:41). UTT's Torre David Gran Horizonte project bought attention to the group's methodology, and this was exhibited at the 2012 Venice Architecture Biennale. Kallipoliti (2013), in a review of the project in the *Journal of Architectural Education*, stated that UTT was a call for the "inevitable reinvention of the architect's identity" (p. 161). Kallipoliti (2013) goes on to summarize the project in the following terms:

> Torre David is our contemporary obligation to face a vertical life raft for thousands of people in reflection of the multidimensional nature of urban space and the unpredictability of life. Documentation and creative projection are not detached and disjunctive realms of operation. (p. 161)

Another notable project by UTT that evidences its design tactics at a project level is the Empower Shack intervention in Khayelitsha South Africa, 35 km from Cape Town (Low, 2018). In this project, UTT proposes a critical alternative to the "Re-Blocking" approach adopted in South African informal settlements. UTT argue that Re-blocking is severely limited as it improves housing without further flexibility for services, open space, or bottom-up economic activity. In contrast, UTT's Empower Shack is created through a prototype housing type that allows for permeability as the circumstances of inhabitants change; a unit may be adapted to become multifamily housing and configured for public interaction and small traders (Low, 2018, p. 93) (Fig. 4.1).

Brillembourg argues for the need to "debunk the myths of marginality," arguing that "informal settlements are not the problem but the solution" (Redento Recio, 2021, 20:38). UTT suggests how informal cities can be explained and theorized. They define the "informal as the condition of complex nonlinear systems, in which boundaries are flexible and patterns overlap intersect and mutate in unexpected ways" (Redento Recio, 2021, 6:31). These patterns can be "termed as emergent patterns of the idea of an array of fields that biologists like to talk about for emergent evolutionary theories" (2021, 6:43). Brillembourg's position on urban theories of complexity is to state that settlements become more static as they grow – countering the idea of a city as an ever-expanding dynamic. Prior to this static point, there is a "moment of flux," that allows the designer to intervene with a degree of agency. Unlike grander visions, UTT adopts an

Fig. 4.1. Torre David, Caracas, Venezuela, 2012 Torre David, Urban Think Tank, (UTT). *Source:* Architectural Design. Brillembourg, A., Klumpner, H., & Kalagas, A. (2015). El Sur global: Future resilient city. *Architectural Design*, *85*(4), 100–105.

interventionist view through architectural techniques and tactics that can be inserted in the medial-scale of rapidly urbanizing cities. This middle-scale approach still clings to the idea that architects have agency and that change can be effected through specific strategies at the project scale. Brillembourg sees architects as critical "mediators" – not as "principal designer" – whose knowledge is important and who should be regarded as moderators of "grassroot" projects (Redento Recio, 2021, 23:35). Thus in contrast to Mehrotra, they assert the importance of architectural agency in relation to informal cities. In doing so, UTT also draws on the ideas of Bernard Rudofsky, Alison and Peter Smithson's CIAM 9 research and John Habraken (PREVI/Lima, 1970).

The work of the humanitarian architect Yasmeen Lari points to an approach that abandons the need to theorize urbanization at the macro-scale. Lari is the cofounder of the Heritage Foundation of Pakistan and the founding chair of the Pakistan chapter of the International Network for Traditional Building, Architecture & Urbanism Congress (INTBAU, 2018). As Pakistan's first female architect, their spatial activism and humanitarian efforts at a microscale include involvement in the construction of one the largest disaster aid shelter programs in the world (Parlor, 2018). Lari also has an interest in vernacular and traditional construction methods through the use of adobe and bamboo. Lari has made these efforts despite Pakistan's patriarchal cultural, religious, and economic contexts. Having completed her architectural training at Oxford Brookes, and their past projects in the early 1980s in Karachi, included the Taj Mahal Hotel (1981), The Finance Trade Centre (1989), and remarkably, given her later career, The Pakistan State Oil House (1991) – all Portmanesque style monsters (Ramzi, 2019). However, after the Baluchistan earthquake in 2005, Yasmin was deeply moved:

> 'At the time I enjoyed using expensive building materials such as large glass panels, polished granite and steel trusses', Lari reflects, 'and perhaps with my present work I am atoning for the damage I caused with my earlier projects', although as she points out, even then she managed to build a few structures using sustainable materials.
>
> (Ramzi, 2019)

A skilled communicator Lari has proposed and developed a "barefoot approach in architecture" and promulgated the principles behind this through the heritage foundation of Pakistan (Heritage Foundation of Pakistan, 2020). Lari's practices include the full range of communication and this includes, education, awareness training, environmental consciousness raising with support from local and international individuals, educational institutions, and organizations. Education is a central focus of Lari's practice, and consequently, INTBAU has promoted a global sustainability design competition and a seminar series regarding climate change planning and rapid urbanization.

There are of course multiple subjectivities that need to be accounted for in the engagement between architects and the informal settlements of the Global South. For instance, Shau is an architectural firm based in Bandung, Indonesia, with strong linkages to the elite architecture schools of the Global North. The firm's founders, Florian Heinzelmann and Daliana Suryawinata, were both educated at the Berlage Institute Rotterdam. Heinzelmann previously worked at UN Studio, and Dalian worked at OMA, MVRDV, and West 8, a multidisciplinary practice with offices in Netherlands, Belgium, and the United States. Shau's current work, design, and research methods can be seen in their SuperKampung studio while working at The Why Factory at TU Delft – the "think tank" vehicle of Winy Maas. The Why Factory had the objective of simulating and visualizing cities of the future (The Why Factory, 2011). In an early project under the SuperKampung label, a design exploration of how a Kampung in Djarkata could be gentrified and reconfigured so that its original inhabitants would not be evicted. The project developed proposals for a "more mixed, dynamic multi-class urban fabric" and "how the most diverse social mixture could be bred out of an existing situation where the rich, medium rich and poor live side-by-side" (The Why Factory, 2011). The aim was to convince policymakers in Djarkata to seek alternative proposals to simply redeveloping Kampungs through gentrification.

A more recent and lauded project of the firm is more prosaic than the grander visions of the SuperKampung. This is the micro libraries project, a prototype for a "low-cost, prefabricated, modular FSC certified timber system" (Chicago Architecture Biennal, 2021). The microlibrary aims to mitigate low levels of literacy in both urban and rural villages in Indonesia. The microlibrary has an elevated book library and storage above a community stage. This upper structure is constructed from steel and concrete with a facade made from 2,000 upcycled ice-cream buckets. These recycled elements form a binary pattern that when decoded says "Books are the windows to the world" (Chicago Architecture Biennal, 2021). The firm was represented at the Chicago Biennale in 2021, shortlisted for the 2019 Aga Khan Award for Architecture and Lafarge Holcim Silver Award Asia-Pacific in 2017, the *FuturArc* magazine designated the firm as "eco-modernists" (Chicago Architecture Biennal, 2021) (Fig. 4.2).

Shau's work points to the paradox of applying eurocentric learned notions of design and aesthetics to the Global South. The rustic aesthetic of the microlibrary is at odds with the firm's European projects, such as the minimalist Embassy of

Fig. 4.2. Microlibrary – Taman Bima, SHAU, Bandung, Indonesia, 2016. *Source:* SHAU. (n.d). *Microlibrary -Taman Bima*. https://www.shau.nl/en/project/53.

the Republic of Indonesia in The Hague completed in 2018. The firm's portfolio of projects in Indonesia appear to exhibit a mix of "acceptable" modernist aesthetics: Eco House Bandung (2019), Microlibrary Selasar (2018), Al Manan Mosque (2018), and a Performing Arts Center (unbuilt, 2017). This aesthetic approach and attendant orientation toward sustainability is also evident in a number of the firm's developer sponsored concept designs for housing proposals. For example, the 600,000 square meter Urban Forest Tangerang (unbuilt, commissioned 2018) and the Cloud Office Park (unbuilt concept 2017). The latter is described as being "clad with a metal mesh made from rings and various disk sizes" (Shau, 2020). More alarmingly, Swarnabumi Residence (ongoing 2015, under construction) is a project with two residential high-rise towers containing over 1,100 small apartment units at Kiaracondong in Bandung (Shau, 2020; Swarnabumi Residence Bandung, 2016).

There are also architects working in the Global South, whom we might designate as outsiders because they are less entwined with architecture's global system of Warlord architects and Megafirms. Either overtly or tacitly, these outsiders are actively engaged with the SDGs. One such architect working as an outsider is Pedro Pacheco in Monterrey, Mexico. Pacheco is the foremost exponent of using waste and recycled materials to build anew in informal communities. An architect trained in both architecture and community development, Pacheco is an Associate Professor at Tec de Monterrey. Their project Casa Rosenda, Diez Casas Para Diez Familias (10 m × 10 m), was constructed for $14,391 USD in 2010. The house is in the informal settlement of the Nuevo Almaguer neighborhood in Guadalupe, Nuevo Leon, and Mexico. The project was built in collaboration with students at Tec de Monterrey, and 40% of the materials of the house were recovered materials. This included fibreglass panels, previously used as concrete molds, vending machine glass for double glazing and left-over wooden planter boxes. Pacheco and his team of Tec de Monterrey architecture students also investigated the use of fridge doors, discarded as damaged seconds from Monterrey's large LG electronics, fridge manufacturing plant which exports fridges to the United States.

Pacheco, in a more recent 2016 project, and using the same techniques as those employed at Casa Rosenda, built Casa Viesca a house for a family in an informal settlement. This was a home for a bricklayer father, the mother a garbage collector, and three children: two sons, one disabled, the other a pallet repairman, and a daughter who was a student. The existing house of the family was built from scrap timber beams and corrugated iron weighed down by bricks. The existing house had no drainage or services. Through a process of consultation with the family, Pacheco built the new home in two phases. Remarkably, the family continued to live in their house while a new steel structure and the first floor was constructed. After this the family lived upstairs while the ground floor was built (P. Pacheco, personal communication, October 2021).

Pacheco's work is not without its theoretical sources in the work of the Catholic thinker Paul Freire whom Pacheco admires. This approach emphasizes the importance of informal education as a way to develop individual awareness and as a "power for transformation." Pacheco cites Freire who says:

> Education for awareness takes its most important role when it becomes a process of cultural action where men and women wake up to the reality of their socio-cultural situation, move beyond their limitations and alienations to which they are subjected and become affirm themselves as conscious subjects and co-creators of their future.
>
> –Paolo Freire. (P. Pacheco, personal communication, October 2021; Freire, 1993)

Freire's project is liberatory, as summarized by Maboloc (2021).

> Education is concerned with liberation. Learning is about the ability of the student to criticise and resist patterns of injustice in society. Students must recognise and react against structural inequality and systemic oppression. Critical pedagogy reflects an awareness of social as well as political issues.
>
> (Maboloc, 2021, p. 1098)

Each of Pacheco's projects emphasizes the need for transdisciplinary coproduction of knowledge between users and others, including experts. Learning and education are central to Pacheco's participatory process: through concepts of playful learning, self-directed learning, and collaborative learning; all these concepts being critical to aligning design processes with critical reflection in action (Pacheco, personal communication, 2021). By linking a liberatory educational theory with, reuse and industrial ecology, Pacheco has created an important model of practice.

Another group of Mexican architects that has begun to develop spatial practices that tacitly aligns with the SDGs as well as the social theories of liberation is Comunal. The group is based in Mexico City, working directly with different Indigenous groups in Mexico. The group was founded in 2015 by two architects Mariana Ordóñez Grajales, a graduate of the Autonomous University of Yucatán and Jesica Amescua Carrera, a graduate of the Ibero-American University. The studio describes itself in the following terms:

> As a team made up of women, we have the social commitment to reflect on the acts of domination and patriarchal violence that are exercised from the architecture on women, their families, communities and territories, with the aim of finding new ways of accompanying social processes related to the right to habitat, from an intercultural stance that rejects racism, colonialism and technocratic visions.
>
> (Comunal, n.d.a)

Like Pacheco, Comunal's approach is collaborative and thus seeks to:

> Collaborate in the improvement of the living conditions and habitability of the rural communities of our country, as well as in the rescue and strengthening of the territorial memory, through integral participatory processes that strengthen local knowledge, autonomy, the exchange of knowledge and resilience, always recognising the inhabitants as at the center of the processes and decision-making.
>
> (Comunal, n.d.b)

One focus of the group is to resist the "extinction" of vernacular housing in Mexico. For example, their project Tequicalco is conceived as a cultural center for the preservation of memory, language, rituals, ideology, politics, and territory of the Nahua People. Community consultation was at the center of this project; for each space, the Nahua People decided upon its function, appropriate construction system, and environmental and site orientation strategy, including the careful preservation of existing trees and alignment with wind and water. Through these means, Comunal hopes to generate autonomy and empowerment in rural Indigenous communities. Comunal's process is collaborative and participatory, acting as facilitators rather than architectural authors. In numerous projects, this involves multiple workshops with the community, and the projects are realized by the community itself – even after Comunal is no longer involved (Comunal, n.d.c).

Comunal's project outcomes are open-ended and not measured by the success of a building but rather by the education and empowerment of the community. Through this way of working, the studio has created a form of design activism that is resistant to colonial and neo-colonial pressures. This approach is the predominant – if not sole – process of the practice and is evident across the projects in the studio's portfolio. This includes work in the Tepetzintan community where the group facilitated "technical construction workshops" employing global bamboo species to develop "the housing exercise in an inclusive way" (Comunal, n.d.d). In Oaxaca the studio worked with the Ixtepecano Committee for the Defense of Life and Territory and the Ixtepec Reconstruction Council, who together with the Universidad de la Tierra de Oaxaca, produced a Self-Build Manual in order to support the "high percentage of families that are rebuilding their homes through self-construction and self-production processes" (Comunal, n.d.e).

Following the 2017 earthquakes in Chiapas and Oaxaca, the studio worked with the Santa María Nativitas Coatlán community, who wanted to rebuild their housing in the manner of their ancestors. For this community:

> ...the way we build our houses is the intelligence of our ancestors and we want to preserve that wisdom.
>
> (Comunal, n.d.f)

The studio worked with this community to develop a local formula for adobe construction for the production of new houses. This entire process involved assessing:

> ...social feasibility, damage census, active participatory investigation of the ways of living, analysis of the failures in the traditional vernacular housing for its improvement, registration of the local formula to make adobes and the processes related to medium-scale production.
>
> (Comunal, n.d.f)

In addition to the above, the studio conducted what it calls "territorial understanding sessions participatory" participatory "cartographies" and mapping that resulted in a produced strategy for the construction of 100,000 pieces of adobe for the rebuilding of houses (Comunal, n.d.f).

The above architects and urbanists – Comunal, Pacheco, Mehrotra, and UTT – can claim a liberation from the orthodox and professional subjectivities of Global North architectural practices. But emerging from Global South, we can also apprehend a range of unique theories and spatial practices, and affordances linked to ideas of decolonizing the Western architectural canon. For example, another architect included in the 2021 Chicago Architecture Biennal is Atelier Masomi, founded by Mariam Kamara. Kamar is based in Niger, having graduated from the University of Washington with a Master of Architecture. In their practice, Atelier Masomi Kamara has completed a number of projects and received numerous accolades (Atelier Masomi, n.d). Notably, the Hikma Religious and Secular Complex, designed by Kamara and Yassi Esmaili, won the 2017 Gold LafargeHolcim Award for Africa and the Middle East and the 2018 Silver Global LafargeHolcim Award for Sustainable Architecture. Mariam Kamara is critical of colonizing architectural aesthetics, as embodied in the Neo Sudanese architecture of French imperialism in Africa. As a result of her own experiences in Architectural education, she argues that the canon of architectural history has focused too much on eurocentric architecture. Speaking at MIT in April of 2021 she states:

> This bias in a way would be fine if it was made clear that it was only one point of view. But the problem is that it is represented as the only point of view a truth about the entire planet. And it kind of presented a vision of place and importance of your Eurocentric architecture and history. And its contribution to history, and understood to be universal and kind of undeniable.
>
> (MIT Architecture, 2021, 15:00)

In order to counter this she argues that architects need to seek:

> ...more complete and more truthful narratives of the world we live in, and how it came to be and you know, alternative ways to look

> at things would probably emerge, you know, on the other end, as in a much more exciting and, frankly, much more interesting world.
>
> (MIT Architecture, 2021, 42:10)

Along similar lines, decolonized spaces are opened up in the research of Huda Tayob, whose research uncovers and recovers the people and infrastructure of incidental and marginal sites across the globe (The Architecture Lobby, 2021). Her research suggests that the canonical archive is part of and implicated with a racial construction of the world where the subaltern voice is made invisible through various mechanisms. However, in the marginal spaces that Tayob studies, she ascribes the possibility of agency, in relation to her work on Somali malls, she states:

> This research points to a wider form and recognition of where knowledge on the urban margins might be located within the spaces of precarity and possibility, and to certain endpoints to certain forms of care and sustenance that are not necessarily reducible to were exhausted by mechanisms of control.
>
> (The Architecture Lobby, 2021, 1:01:58)

Her research argues that informal sites:

> ...are not simply ad hoc incidental sites they have a history, which is linked to formal infrastructure refugee policies, global capital colonial planning and state dispossession, and they are tied into the ongoing racialisation of migration and marginalisation of these networks and practices, which continues into the present across the Global South and North. (1:00 52)

As Tayob posits, the prevailing architectural canon reflects that the cities of the Global South are invisible because they exhibit a lack, and that the "disciplinary archive is deeply implicated with the racial constructions of the world." (1:01:46).

Counterspace studio, based in Johannesburg, exemplifies how different spatial practices aligned with an expanded theoretical lens can be activated. Counterspace was cofounded by Sumayya Vally and describes itself as a collaborative architecture and research firm that occupies a space "between the functional and the speculative; pedagogy and praxis" (Counterspace, 2021a). In many respects, the studio is exploring and developing new collaborative mechanisms for architects working in both the Global South and North. Notably, Sumayya was the designer of the 2020/2021 Serpentine Pavilion in London, and a part of this project was aligned to "Support Structures for Support Structures," a program initiated by the Serpentine Gallery that provides fellowships for up to 10 artists (Serpentine Galleries, 2021).

The program supported spatial practitioners working across a range of fields at once, including art, politics, and community practice. This range of practice is often not supported by traditional grant institutions. For example, the program funded Barby Asante an artist-curator and "healer in training" whose practice is concerned with "the politics of place, space and the ever-present histories and legacies of slavery and colonialism" (Counterspace, 2021b). Asante's 2017 performance work *Intimacy and Distance* at the 2017 Venice Art Biennale explores ideas related to:

> ...post-colonial identities, migrations and journeys, political, social and cultural agency, kinship amongst women of colour and everyday methods and strategies of decolonisation.
>
> (Barby Asante, 2021)

Another recipient of the Serpentine fellowship is the Nawi Collective (2020), who self-describe themselves as a "London-based black women & femmes vocal collective. Singing to reclaim our time. Singing for our ancestors" (Nawi Collective, 2020). This group employs song as a

> ...portal to preserve their traditions, praise their ancestors and connect and commune with each other. The spaces they create together when they sing offer powerful healing amidst a world that does not care for or foster the innate power, love and radical possibility they each hold.
>
> (Counterspace, 2021b)

Through Counterspace's collaborations across the creative humanities, an expanded range of spatial practices is highlighted, as well as all this, subaltern cities are engaged with. For example, Vally's practice is predicated on a critique of the central canon of architectural education. As Vally states:

> ...with things that are beyond the formal archives so working with recipes, working with songs and stories and so on, and really pushing these and treating them with, you know as much respect as we do work in material that's in the formal archive or that has been formally mapped and documented I think is really important as a contribution to making a visible archive of course but also as inspiration for new design form.
>
> (Cooper Union, 2020, 58:30)

If this polemic appears to be removed from a focus on the SDGs and the climate emergency, then Vally points to the need to:

> ...think through and respond to the inextricable connections between history forces of labor, race and class struggles, capitalism, toxicity and climate change.
>
> (Cooper Union, 2020, 6:34)

This leads to an approach to practice, as evident in her design studio teaching at the Graduate School of Architecture Johannesburg, where Vally explores:

> ...design languages for cultural hybridity and identity, gender, race, politics, orders storytelling narrative myth, performance, the way with the embodied the overlooked.
>
> (Cooper Union, 2020, 42:27)

The resultant expanded field of architectural theory and practice is identified in the seven selected points taken from her 24-point polemical love letter published in the *Architectural Review* in September 2020, with points including:

(1) you will soon fall in love with gold, kitsch, supernatural ideas,
(2) the tenacity of indigenous plants and indigenous ceremonies and practices – all the magic that is Joburg. There is another canon here.
(3) Ingest atmospheres – learn how to read and feel color, dust, mist, the phases of the moon. There is another canon here.
(4) Look at these things deeply.
(5) Feel them, absorb them.
(6) You will soon develop a mistrust for the historical record. Listen to that.
(7) Look so deeply at what is present that you notice the silences and the absences too. There is yet another canon here, in these silences and absences. (Vally, 2020, p. 39).

And through these practices, collaborations, teaching, and polemics, Counterspace's work indicates the possibilities of less circumscribed architectural histories and narratives, an expanded field of subjectivities that will then give space to a greater consideration of the SDGs and climate by the architects of the Global North.

2021 Venice Architecture Biennale, Global South Cities and Climate Change

The curation of the 2021 Venice Architecture Biennale indicates the extent to which architects have addressed the, the SDGs, the cities of the Global South and the environmental politics of architects. The 2021 Biennale was curated by Hashim Sarkis, an architect and Dean of the School of Architecture and Planning at MIT. The theme for the entire exhibition is entitled, "How will we live together?" (Biennale Architettura, 2021a). Sarkis saw this question as an open question that would enable the structuring of the exhibition around a number of subthemes and concerns. For Sarkis this overall theme speaks to different issues. As a rhetorical question the "How" is intended to address "practical approaches and concrete solutions"; the "Will" is focused on the future and "seeking vision and determination, drawing from the power of the architectural imaginary," the "We" is concerned with being "inclusive of other peoples, of other species" and an "empathetic" approach to architecture, and finally the "Live" is defined as to "flourish" and "thrive"; and together this implies collective and universal values (Biennale Architettura, 2021a). This question is explicated in the following manner:

> It is indeed as much a social and political question as a spatial one. More recently, rapidly changing social norms, the political polarisation between left and right, climate change, and the growing gap between labor and capital are making this question more urgently relevant and at different scales than before.

The solution to these dilemmas is then put forward:

> In parallel, the weakness of the political models being proposed today compels us to put space first and, perhaps like Aristotle, look at the way architecture shapes inhabitation in order to imagine potential models for how we could live together.
>
> (Biennale Architettura, 2021b)

But in adopting the abstract, if not ambiguous idea of space, alongside suggesting the necessity for an apolitical viewpoint, to answer their own overarching question, How will we live together? Sarkis brings plurality to the rescue:

> Today, unlike with previous ideologically-driven generations, there seems to be a consensus that there is no single source from which such an answer can come. The plurality of sources and diversity of answers will only enrich our living together, not impede it.
>
> (Biennale Architettura, 2021b)

Following on from the initial question and its rhetorical parsing. Sarkis subdivided the Biennale into three more subthemes: Among Diverse Beings, As New Households, and As Emerging Communities – the exhibits associated with these themes were exhibited in Venice's Corderie 316 m long former rope workshops of the Arsenale at the center of the Venetian Republic's maritime power. To complicate matters, if not to ensure all bases were covered, each of these themes described further subthemes: Among Diverse Beings included: Designing For New Bodies and Living With Other Beings. As New Households included: Catering To New Demographics, Inhabiting New Tectonics and Living Apart Together. As Emerging Communities encompassed the further concerns of Appealing To Civicness, Re-Equipping Society Coming and then a section entitled coming Together In Venice Co-Habitats (Biennale Architettura, 2021a).

Just beyond the Corderie, just inside the Artiglierie (exhibits C3–C12), Sarkis included a section entitled "How We Do Live Together In [city/place]." It was in this section that the cities of the Global South were highlighted. These exhibits included architects who had been invited to exhibit from a range of countries, cities and places: Nigeria, Egypt, Mexico, Addis Ababa São Paul, Al Azraq

Camp, Beirut, Rio De Janeiro Prishtina, Hong Kong, and somewhat surprisingly New York (Biennale Architettura, 2021a). In the Central Pavilion of the Biennale, Sarkis organized the exhibits around the following two larger themes and corresponding subthemes (Theme: Sub-Theme). Across Borders: Protecting Global Commons, Transcending The Urban-Rural Divide, Linking The Levant, Seeking Refuge and Re-Sourcing Resources. And then As One Planet: Making Worlds, Designing For Climate Change and Networking Space (Biennale Architettura, 2021a).

The Designing for Climate Change exhibit included the work of three projects. Each of these projects did not suggest how architects might develop new design methods or practices in order to design for climate change. Each project was more didactic and conceptual. Firstly, Melting Landscape by Kei Kaihoh Architects was a conceptual art piece that explored Yukimura, a Japanese vernacular snow storage method for food preservation on Japan's mountains. The intent of the installation was to explore new ways of using snow and to understand new technologies that might be applied to it (Biennale Channel, 2021b; Kei Kaihoh Architects, 2021). This project advanced the idea of design innovation, thus seeing the climate emergency as an opportunity to adapt traditional practices. The second project included in Sarkis's climate change category was by Design Earth. This exhibit, *The Planet After Geoengineering*, presented a dystopian future where humankind, as well as the Earth's atmosphere which, could only survive as a result of geoengineering. Design Earth presented five "geostories," a series of dystopian moral tales for the Anthropocene. These narratives were designed to make climate engineering visible. Design Earth project clearly see the architectural project as a means of communication – a speculative practice of public communication – that would create alternative and dystopian worlds engendered by the climate crisis (Biennale Channel, 2021a; Design Earth, 2021). The third project in this section was a collaboration between OOZE and Marjetica Potrč. These collaborators exhibited a miniature prototype of a public art project Future Island. Future Island is a landscape urbanism art project, 44.5 m × 16.9 m in size, sponsored by the Swedish Arts Agency and Stockholm University, located at the Albano campus of the University, with a completion date of 2022. The project exhibited at Venice was a miniature prototype of the island; the final project is an island of stones, a landscape that will evolve over the next 100 years; one half of the island will demonstrate the impact of an earth that is overheating (Biennale Channel, 2021c; OOZE e Marjetcia Potrc, 2021).

The above curation by Sarkis should only be viewed as a classification system, a taxonomic regime, expressing a number of architectural tendencies that do not radically question the prevailing architectural canon. Nonetheless, a dichotomy can be discerned in this curatorial regime. Interestingly, the first two themes "Among Diverse Beings" and "As New Households" appeal to the polemics and experiments of those architects concerned with the biodigital. Whereas, "As Emerging Communities" appears to highlight architects interested in bottom-up notions of the civic, social processes and the cities and the places of the Global South. While a number of projects the projects selected by Sarkis discussed issues related to the climate emergency, many were overtly conceptual, steeped in the deliria of climate imaginaries, bluntly dystopian, or symptomatic of a kind of climate deafness – as evidenced by the three projects in the Designing for Climate Change section. However, in this curatorial system, the rapidly urbanization cities of the Global South that only appear as peripheral moments; and for many, if not all the exhibiting architects at Venice in 2021, each of UN-Habitat's SDG's are apparitions like the Venetian galleon's that now haunt the Arsenale.

Warlord Architects and the Global South

Warlord architects, as I have written elsewhere, is a more aggressive moniker for the class of architects often designated as star architects. This is because Warlord as a name better describes the violent and ecocidal activities of these architects. These activities are driven by a desire to increase their own agency through various media channels. But also, within the ranks of Warlord architects we can see gendered masculinities, business as usual models, a disinterest in professional reform, and an attachment to the canon of Anglo-centric architecture (Raisbeck, 2019, p. 66). Warlords have at times pursued projects in various developing cities and contexts. Often these engagements are portrayed as a kind of enlightened philanthropy. As I wrote elsewhere: "But what is given by warlords through this philanthropy are the 'jewels' of their architectural production rather than money" (Raisbeck, 2019, p. 83). These engagements, for instance, Koolhaas's engagements in Lagos, suggest the way in which warlord architects embed themselves, in a "philanthropic discourse" of global urbanism (Raisbeck, 2019, p. 83).

Rasmussen expands this point by arguing that celebrities linked to particular global development initiatives curtail and distract any gaze from the realities of living in an informal settlement (2019, p. 83). A small body of research has grown ups around the phenomena of celebrity advocacy and international development. An even smaller body of research has focused on celebrity architects and their projects in the Global South. One exception is Bianco (2018), who studied three projects in the Global South by three Pritzker Prize–winning architects: Renzo Piano's City Gate Project in Valletta Malta, Foster and Partners Master plan for Dharavi in Mumbai India, and Rem Koolhaas's Development of Beirut's Dalieh coast in Beirut Lebanon (Bianco, 2018, p. 96). Bianco (2018) noted that the opposition from the public prevented the latter two of these projects from going ahead; concluding the importance of the participation of civil society in the development of the projects. Bianco provides a long quote from, and 2014 interview in the *European Magazine* from Norman Foster regarding the Mumbai Dharavi project:

> We developed a comprehensive plan to improve the quality of life for all living there, which was based around the existing balance between spaces for living and working, yet introduced new public facilities and infrastructure. While we sadly have not had the opportunity to implement our master plan, this work has been a valuable reference for potential future projects. Significantly, it pointed the way to solutions in which the community would be

> respected and the quality of amenities transformed. This is a radical alternative to the traditional approach of bulldozing, uprooting the social structure, and starting afresh – a policy which has so far failed.
>
> (Foster, 2014)

Foster's tone of paternalism belies that in order to fund community infrastructure, the inhabitants would be rehoused in high-rise housing, which would then be sold "for public sale." The $1.2B project was to be financed through a Public–Private Partnership mechanism (Bianco, 2018, p. 97). Bianco (2018), in documenting the community groups that opposed each of the developments, concluded that:

> The language of architects on architecture is rhetorical, a language ignored by the public. The approach of architects is paternalistic, whereas civil society demands a bottom-up approach which considers living realities, values, and perceptions of the public. (p. 98)

Foster's paternalism can also be seen writ large in the global master planning vision of Bjarke Ingels. Armed with planetary ambitions, and the self-confidence of architectural design thinking, Ingels has no need for the niceties of the SDGs. This disregard is evident in Bjarke Ingels's somewhat bizarre ambition to master plan the planet as a way to solve the climate emergency. Featured in an interview in *Time* magazine in October 2020, Bjarke's plan quickly gained media traction in *Dezeen*, *ArchDaily*, and *The Guardian* (Nugen, 2020). Ingels struck on the ad hoc idea of master planning the planet following a project for Akio Toyoda of Toyota at the base of Mount Fuji for an experimental city. Prior to the *Time* article, in a lecture at Columbia GSAPP in 2020, Ingels outlines a mish-mash of futurist technologies: robotics, smart homes, connectivity through AI, multigeneration neural assisted living, hydrogen-powered infrastructure, as well as carbon capture (Columbia GSAPP, 2020, 52:51). This futurism is also evident in other projects including Hyperloop One, a spaceport for the Norwegian government in the Arctic Circle, and a project for a human colony on Mars. In the Columbia lecture Ingels went on to question the ability of both scientists and politicians to mitigate climate change. In Ingel's account climate scientists:

> ...because they are a majority of academics implies, that they're very good at science and academic accuracy but not so much at entrepreneurship and action. (1:07:30)

Moreover, politicians are bound by election cycles who:

> ...are maybe not so good at something that requires a 50 year, or 100 year commitment, because they have election cycles of four or eight years. (1:07:30)

And then the heroic architects arrive to rescue the planet:

> So what we thought what if we because architects, maybe we make master plans for buildings for city blocks for neighbourhoods for cities or regions even for countries, why not make a master plan for the planet. (1:07:30)

Ingel's global approach evokes, yet again and perhaps not surprisingly, the American inventor and engineer Buckminster Fuller. At the sixth International Union of Architects (IUA) Conference in London in 1961, Fuller proposed that the architecture schools of the world "undertake a continuing study of the chemical and energy resources of the world" (McHale, 1964). The echoes of Fuller in Ingel's work suggest an ad hoc recycling of old tropes and philosophies in architectural rhetoric: a planetary holism, interconnected systems, and fundamental patterns of nature revealed by science – not to mention a paucity of realistic ideas. All these elements can only be realized via the actions of the architect in order to effect a faux planetary holism. While Ingels appears to be dimly aware of the SDGs, with their complex interdependencies, his global and diagrammatic worlding of the climate emergency ignores climate impacts on the Global South. This is architecture on a global scale with no regard for the climate emergency in the cities of the Global South, and perhaps the real historical echo here – as with Ecologic – is the fascist ambitions and gigantism of Albert Speer.

The SDG's and Coproduction

To proceed, we must now ask how should the SDGs and the NUA be implemented within architectural discourse and practice? Valencia et al. (2019), in an extensive study of seven cities (Buenos Aires, Cape Town, Gothenburg, Kisumu, Malmö, Sheffield, and Shimla), identify a number of critical implementation factors. These authors argue that there are five interrelated SDG adaption or localization factors that need consideration: delimitation of urban boundaries; inclusively identifying the full range of actors; the synergies and trade-offs between SDGs as some may require more focus in a given location; determining the relevant data gathering and indicators to measure sustainability progress plus integrated governance that addresses multiple dimensions of sustainability (Valencia et al., 2019, p. 4). They assert that the "integrated governance approach" requires clear data management that is not simply managerial benchmarking – or global imaginaries, we might add – but is self-reflexive enough to identify gaps in implementation (Valencia et al., 2019, pp. 19–20). As they note:

> Political will, coherent governance, and strong formal partnerships between the public sector, private sector and civil society actors are key ingredients in achieving goals such as the SDGs. (p. 20)

In other work Veland et al. (2021) argue that the SDG goals are susceptible to being hijacked by "habitual tendencies" of managerialism. This habitual tendency is clearly evident in the work of seemingly disparate architects such as Ingles and Ecologic. A limiting tendency that is characterized by the architect's desire to control, either territories or technologies, and produce so-called solutions. Interestingly, Veland et al. (2021) propose a counter to this tendency through the creative practices of the environmental humanities. They see utilizing music and emotion as a way to unlock the potential of the SDGs. While this may sound unconventional, this approach highlights the potential of "co-creative spaces" in SDG implementation (p. 11). Veland et al. (2021) argue this idea in the following way:

> Given the shortcomings of target- and indicator-based management of societal transformation, the room for filling each goal with emergent and contextual meaning, and the infinite complexity of ultimate goal integration, we further explore opening co-creative spaces for SDG implementation. (p. 11)

For architects, cocreative spaces is arguably an easily grasped concept and also leads to the idea of coproduction; despite the tendency of some architects to foreclose this manner of practice; or view coproduction as simply being about integrating knowledge with other AEC consultants. It is incorrect for architects to suggest that the coproduction of design knowledge has no place in sustainability transitions, or any design process for that matter – as Ingels appears to do in his rejection of climate scientists. It is obvious that at both the firm and project level, architects coproduce design knowledge across disciplines and with different actors, entities and artifacts. The coproduction concept originated in research that examined how public and public–private services were coproduced between government entities and citizens (Ostrom, 1996).

Coproduction, as introduced in Chapter 2, has since the late 1990s, also evolved in climate studies. We now return in more detail to Bremer and Meisch (2017) and their identification of two prevailing approaches regarding coproduction and climate research. The point here is that their articulation of coproduction highlights a kind of paucity in architectural theory and practice. In their bibliometrics study, as noted previously, the two approaches are designated as the normative or the descriptive. The normative approach, as seen in the next chapter in relation to sustainability metrics and rating tools, is described as:

> The normative approach aims to develop and articulate guidelines or norms on how various actors, including government, climate scientists and stakeholder citizens should define and co-produce knowledge.

In this approach knowledge has valuable because of its utility:

> In the normative approach this the emphasis is on co-producing, through interactive and interactive processes, of useable

> knowledge and norms for policy implementation. (Bremer & Meisch, 2017, p. 2)

In contrast, the descriptive approach seeks to avoid managerial interventions and aims to describe and map:

> …how the co-production of various artifacts, settings, knowledge, social orders, and power relationships happens. We call this area descriptive because it mainly uses the co-production idiom for interpreting the shifting relationships between science, society, and nature – including around climate change – rather than intervening to actively change these relationships.
> (Bremer & Meisch, 2017, p. 2)

In this descriptive approach, various authors such as Latour (1993) and Jasanoff (2004) go further, emphasizing the cyclical making and remaking of knowledge between society, technology, science, and nature. In this more situated approach, because of its focus on artifacts and settings, any analysis has an emphasis on interpreting and describing dynamic relationships, particularly those related to power asymmetries. This is therefore work that examines technoscientific objects, boundaries, framings, relationships, and cultural practices between society, science, and nature. Bremer and Meisch (2017) extend the above two distinctions to create a nuanced model, through a series of eight lenses, for thinking about coproduction in climate research – and architectural contexts for that matter. In this model four normative points – Extended Science, Iterative Action, Social Learning, and Empowerment – and associated questions, are explained as:

> Extended science: How can we integrate non-scientists as co-investigators in extended modes of climate science? Iterative interaction: How can we promote consultative interaction between climate science providers and users, for more useful information? Social learning: How can we facilitate social learning about climate issues? and finally Empowerment: How can we empower traditional ecological knowledge systems for governance? (2017, pp. 6–8)

In contrast, the four descriptive lenses are explained as:

> …a constitutive lense; How does climate change shape our representation of nature, and ourselves? Interactional lense; How does climate science evolve with institutions and wider social, political and economic systems? Institutional; How can we build adaptive capacity in governance institutions? Public services; How can government agencies and citizens jointly produce public services? (2017, pp. 8–11)

At the broader scale the normative versus descriptive split is discernible in sustainability research in architecture. This divide is evident in the claims of green researchers – in their academies – aligned with technical solutions, policy prescriptions, and certification methods versus those architectural researchers interested in understanding more situated social-technical and socio-nature framings of architecture and the climate. Arguably the former are less aligned, have no concern, or no knowledge of the decolonizing practices of architects working in the Global South. Certainly, few architectural researchers have mapped the intersections between power, architecture's spatial practices, and climate change. To overcome such impasse's Bremer and Meisch (2017) argue that there are other ways to map coproduction in relation to climate science; they argue for a "a richer picture of the terrain of climate change co-production" (p. 2). They see coproduction as "a complex meeting place where several different academic traditions and practices converge, overlap, affect each other, come into conflict, or cooperate toward describing and effecting co-production" (2017, p. 2).

For architects, this model of Bremer and Meisch (2017), as hinted above suggests much else as well; not the least the problematics of both the normative and descriptive debates in architectural discourse. For architects the above lenss might evoke a number of trajectories. If embedded in architectural theory coproduction suggests how architects might develop new modalities of practice that abandons the reified tropes of warlord architects. This new modality of practice would see informal cities and settlements, with their rapid urbanization, as places with both educative and liberatory capacities. In this modality these cities would be seen fundamentally as something from which architects gaze out from rather than something that they gaze over and into with their filleting knives of neoliberal urbanism.

Conclusion

The predilection of architects to enunciate global issues and design concepts leads to a dilemma. On the one hand the global scale is necessary for apprehending the nature of the climate emergency. On the other hand, architecture itself, as a geographic scale, as a scale of consideration that enables particular kinds of sustainability transitions, is missing in the SDG and NUA research and policy discourse. The only recourse for architects then is the production of climate imaginaries, both in theory and practice, that draws on the imaginary episteme and logics of a benevolent future Anthropocene – as explored in the previous chapter. Alternatively, architects might begin to develop new theoretical concerns that begin to decolonize their own canon and histories. These new theoretical concerns would be hopefully grounded in the cities and settlements of the Global South.

We may then conclude this chapter by pursuing the dichotomy between the enlightened philanthropy and global aspirations of Foster, Koolhaas, and Ingels against the architects grounded in the context of informal cities. This comparison points to different engagements with the SDGs and different modalities of coproduction. For Warlord architects the framework and system of governance related to the SDGs is of no interest; except as an empty rationale for their own solipsistic tendencies. The Warlord architect has no need to be involved in the UN-Habitat system – except as a media opportunity – or indeed to be engaged with civil society, especially when they can master plan the planet. In contrast, the architects discussed in this chapter, working in cities in the Global South, indicate how architects might begin to coproduce knowledge in new and expanded fields of practice and design. This will require a requestioning of the monolithic and eurocentric architectural canon and its imbrication in systems of power, privilege, and gatekeeping in architectural education. And it is in architectural education and its modes of production – for example, in the global theater of the invited guest lecture – where the elision and tensions, between center and periphery, in the production of architectural knowledge are most evident. To finish the chapter, two questions can be posed more bluntly: Are sustainability transitions only produced through centers of privileged knowledge and expertise? Or are sustainability transitions best coproduced through an engagement with the cities of the Global South?

References

Africapolis. (2021a). About. Retrieved from https://africapolis.org/about

Africapolis. (2021b). Beyond large cities. Retrieved from https://africapolis.org/en/about/beyond-large-cities

Asante, B. (2021). Intimacy and distance. Retrieved from https://www.barbyasante.com/work/intimacy-and-distance

Bianco, L. (2018). Architecture, values and perception: Between rhetoric and reality. *Frontiers of Architectural Research*, *7*(1), 92–99.

Biennale Architettura. (2021a). Hashim Sarkis how will we live together – Biennale Architettura 2021. [brochure]. Retrieved from https://static.labiennale.org/files/architettura/Documenti/brochure-2021-en6.pdf

Biennale Architettura. (2021b). Statement by Hashim Sarkis Curator of the 17th international architecture exhibition. Retrieved from https://www.labiennale.org/en/architecture/2021/statement-hashim-sarkis

Biennale Channel. (2021a, April 30). Biennale Architettura 2021 – Sneak Peek: Design earth [video]. YouTube. Retrieved from https://www.youtube.com/watch?v=CSP1j0s2iI8&t=10s

Biennale Channel. (2021c, August 23). Biennale Architettura 2021 – OOZE and Marjetica Potrč. Retrieved from https://www.youtube.com/watch?v=feHV2Z4GV28&t=5s

Biennale Channel. (2021b, August 23). Biennale Architettura 2021 – Kei Kaihoh architects [video]. YouTube. Retrieved from https://www.youtube.com/watch?v=t6AnfhTqg24&t=14s retrieved 221121

Bremer, S., & Meisch, S. (2017). Co-production in climate change research: Reviewing different perspectives. *Wiley Interdisciplinary Reviews: Climate Change*, *8*(6), e482. doi:10.1002/wcc.482

Brillembourg, A., Feireiss, K., & Klumpner, H. (Eds.). (2005). *Informal city: Caracas case*. New York, NY: Prestel.

Chicago Architecture Biennale. (2021). Shau. Retrieved from https://chicagoarchitecturebiennial.org/current_edition/contributors/shau

Columbia GSAPP. (2020, April 25). Bjarke Ingels [video]. YouTube. Retrieved from https://www.youtube.com/watch?v=4Z82m-PxDO0

Comunal. (n.d.a). Who we are. Retrieved from https://www.comunaltaller.com/nosotras

Comunal. (n.d.b). Mission. Retrieved from https://www.comunaltaller.com/mision

Comunal. (n.d.c). Tequiocalco + Communal working group. Retrieved from https://www.comunaltaller.com/tequiocalco

Comunal. (n.d.d). *Social production of housing: Exercise 01*. Retrieved from https://www.comunaltaller.com/ejercicio01

Comunal. (n.d.e). self-build manual. Retrieved from https://www.comunaltaller.com/manualautoconstruccione

Comunal. (n.d.f). Territorial understanding: Sierra Mixe de Oaxaca. Retrieved from https://www.comunaltaller.com/entendimientoterritorial

Cooper Union. (2020, November 26). Visiting lecture—Sumayya Vally, Counterspace studio: Counterpoint [video]. YouTube. Retrieved from https://www.youtube.com/watch?v=ep_bVF3hkX4&list=PL-WfLn9ekLD8XvKZmZRj1JM7K81tE2piA&index=1

Counterspace. (2021a). About Counterspace. Retrieved from https://counterspace-studio.com/about

Counterspace. (2021b). Support our support structures. Retrieved from https://counterspace-studio.com/projects/support-structures-for-support-structures

Das, S. M. (2021, 31 July). With climate change, nature will command us: Sustainable architects must respect that. Times Evoke. Retrieved from https://timesofindia.indiatimes.com/with-climate-change-nature-will-command-us-sustainable-architects-must-respect-that/articleshow/84903471.cms

Design Earth. (2021). The planet after geoengineering. Retrieved from https://www.labiennale.org/en/architecture/2021/one-planet/design-earth

Dovey, K., Shafique, T., van Oostrum, M., & Chatterjee, I. (2021). Informal settlement is not a euphemism for 'Slum': What's at Stake beyond the language? *International Development Planning Review*, *43*(2). doi:10.3828/idpr.2020.14

Dovey, K., van Oostrum, M., Chatterjee, I., & Shafique, T. (2020). Towards a morphogenesis of informal settlements. *Habitat International*, *104*(102240). doi:10.1016/j.habitatint.2020.102240

Foster, N. (2014, October 16). Architecture is an expression of values. The European. Retrieved from http://www.theeuropean-magazine.com/norman-foster/retrieved 120921

Freire, P. (1993). *Pedagogy of the oppressed.* New York, NY: Continuum.

Gaziulusoy, I., & Erdoğan Öztekin, E. (2019). Design for sustainability transitions: Origins, attitudes and future directions. *Sustainability*, *11*(13), 3601. doi:10.3390/su11133601

Heritage Foundation of Pakistan. (2020). About us. Heritage Pakistan. Retrieved from http://www.heritagefoundationpak.org/Page/1307/about-us-registration-mission-vision-board-of-directors-hf-offices-registration-heritage-foundation

INTBAU. (2018). Chapters Pakistan. Retrieved from https://www.intbau.org/chapters/pakistan/

Jasanoff, S. (Ed.). (2004). *States of knowledge: The co-production of science and social order*. Routledge.

Kallipoliti, L. (2013). Torre David/Gran Horizonte. *Journal of Architectural Education*, *67*(1), 159–161. doi:10.1080/10464883.2013.767137

Kei Kaihoh Architects. (2021). Melting landscape. Retrieved from https://www.labiennale.org/en/architecture/2021/one-planet/kei-kaihoh-architects

Köhler, J., Geels, F. W., Kern, F., Markard, J., Onsongo, E., Wieczorek, A., … Wells, P. (2019). An Agenda for sustainability transitions research: State of the art and future directions. *Environmental Innovation and Societal Transitions*, *31*, 1–32. doi:10.1016/j.eist.2019.01.004

Kvan, T. (2020). Data-informed design: A call for theory. *Architectural Design*, *90*(3), 26–31. doi:10.1002/ad.2564

Latour, B. (1993). *We have never been modern*. New York, NY: Harvard University Press.

Low, I. (2018, May 14). Family business: Empower Shack in Khayelitsha near Cape Town, South Africa by urban-think tank and ETH Zurich. *The Architectural Review*. Retrieved from https://www.architectural-review.com/author/iain-low-2

Maboloc, C. R. (2021). Deep thinking or resistance? On finding a middle ground between Paolo Freire's critical pedagogy and John Dewey's pragmatism. *Philosophia*, *49*, 1097–1108. doi:10.1007/s11406-020-00292-5

Masomi, A. (n.d). Our team: Atelier Masomi. Retrieved from http://www.ateliermasomi.com/team

McHale, J. (1964). The geoscope. *Architectural Design*, *34*(Dec), 632.

Mehrotra, R. (2020). Urban design: Imaginations beyond architecture. In F. Carré & M. Chen (Eds.), *The informal economy revisited* (pp. 132–135). New York, NY: Routledge.

Mehrotra, R., & Vera, F. (2017). Ephemeral urbanism: Looking at extreme temporalities. In T. Hass & H. Westlund (Eds.), *The post-urban world* (pp. 44–55). London: Routledge.

MIT Architecture. (2021, April 24). Mariam Kamara | how we narrate our yesterday determines how we imagine the future of architecture. Spring 2021 Lecture Series [video]. YouTube. Retrieved from https://www.youtube.com/watch?v=kRUJ6ZEVzf4

Nawi Collective. [@nawicollective). (2020, February 20). [Instagram Post]. Retrieved from https://www.instagram.com/p/B8yUHR6hRke/

Nugen, C. (2020, October 21). The climate is breaking down. Architect Bjarke Ingels has a masterplan for that. *Time*. Retrieved from https://time.com/collection/great-reset/5900743/bjarke-ingels-climate-change-architecture

Ostrom, E. (1996). Crossing the great divide: Co-production, synergy, and development. *World Development*, *24*(6), 1073–1087.

OOZE E Marjetica Potrč. (2021). Future Island in Venice: The time of stone. La Biennale di Venezia. Retrieved from https://www.labiennale.org/en/architecture/2021/one-planet/ooze-e-marjetica-potr%C4%8D

Parlour. (2018, July 27). The architecture of humanity. Parlour: Women, equity, architecture. Retrieved from https://archiparlour.org/the-architecture-of-humanity

PREVI/Lima Low Cost Housing Project. (1970, April). *Architectural Design*, *40/4*, 187–205.

Raisbeck, P. (2019). *Architecture as a global system: Scavengers, tribes, Warlords and Megafirms*. Bingley: Emerald Publishing Limited.

Ramzi, S. (2019, September 9). Retrospective: Yasmeen Lari. *The Architects Review*. Retrieved from https://www.architectural-review.com/buildings/retrospective-yasmeen-lari

Redento Recio. (2021, September 1). *The paradox of informal urbanism, Alfredo Brillembourg*. The University of Melbourne. *InfUr Webinar Series*, [video]. YouTube. Retrieved from https://youtu.be/RagCq7vfOSM

Redento Recio. (2021a, November 3). Idioms of informality. Fran Tonkiss The University of Melbourne. *InfUr Webinar Series*, [video]. YouTube. Retrieved from https://www.youtube.com/watch?v=il6HdpmB_10&t=298s

RMA Architects. (n.d.a). Rahul Mehrotra. Retrieved from http://rmaarchitects.com/rahul-mehrotra

RMA Architects. (n.d.b). Exhibitions. Retrieved from http://rmaarchitects.com/category/exhibitions/

Roy, A. (2011). Slumdog cities: Rethinking Subaltern urbanism. *International Journal of Urban and Regional Research*, *35*(2), 223–238.

Serpentine Galleries. (2021). Support structures for support structures. Retrieved from https://www.serpentinegalleries.org/whats-on/support-structures-for-support-structures/#:~:text=Support%20Structures%20for%20Support%20Structures%20is%20a%20fellowship%20programme%20initiated,10%2C000%20to%20develop%20creative%20ideas

Shau. (2020). Shau Projects. Retrieved from https://shau.nl/en/projects

Stiftel, B. (2021). Planners and the new urban Agenda: Will we lead the agenda, or will the agenda lead us? *TPR: Town Planning Review*, *92*(4), 421–441. doi:10.3828/tpr.2021.13

Swarnabumi Residence Bandung. (2016, September 20). Apartemen Swarnabumi residence Bandung – 3D impression [video]. YouTube. Retrieved from https://www.youtube.com/watch?v=-dz3vo2OEF0

The Architecture Lobby. (2021, June 20). C-A-P-I-T-A-L-I-S-M ABC school speaker session—Huda Tayob, black markets [video]. YouTube. Retrieved from https://www.youtube.com/watch?v=FElmIgvS1Vk

The Why Factory. (2011). Superkampung envisioning a multi-class society in Jakarta. Retrieved from https://thewhyfactory.com/education/superkampung

UN-Habitat. (2020). Climate change vulnerability and Risk – A guide for community assessments, action planning and implementation. United Nations Human Settlements Programme. Retrieved from https://unhabitat.org/sites/default/files/2020/05/climatechange_vulnerabilityandriskguide.pdf

UN-Habitat III. (2015, May 31). Habitat Iii issue papers 22 – Informal settlements. In United Nations conference on housing and sustainable development, October 2016. Retrieved from https://unhabitat.org/sites/default/files/download-manager-files/Habitat-III-Issue-Paper-22_Informal-Settlements-2.0%20%282%29.pdf

United Nations. (Dept. Economic and Social Affairs). (2017). *Transforming our world: The 2030 agenda for sustainable development*. United Nations. Retrieved from https://sdgs.un.org/publications/transforming-our-world-2030-agenda-sustainable-development-17981

Valencia, S. C., Simon, D., Croese, S., Nordqvist, J., Oloko, M., Sharma, T., … Versace, I. (2019). Adapting the sustainable development goals and the new urban Agenda to the city level: Initial reflections from a comparative research project. *International Journal of Urban Sustainable Development*, *11*(1), 4–23.

Vally, S. (2020, September 14). Sumayya Vally: Letter to a young architect. *The Architectural Review*, 38–39.

Veland, S., Gram-Hanssen, I., Maggs, D., & Lynch, A. H. (2021). Can the sustainable development goals Harness the means and the manner of transformation? *Sustainability Science*. doi:10.1007/s11625-021-01032-8

Vera, F., & Sordi, J. (Eds.). (2015). *Andrea Branzi: From radical design to post-environmentalism*. Providencia: ARQ Ediciones.

Vidler, A. (2008). Whatever happened to ecology? John McHale and the Bucky Fuller revival. *Log*, *13/14*, 139–146.

Chapter 5

Green Architecture and the Good Anthropocene

Introduction

Aldo Rossi's book *Architecture and the City* serves as a useful preface to this chapter on how architects have engaged with the city in relation to the climate emergency (Rossi, 1982). Following the previous chapters, Rossi's book points to a different and perhaps refreshing perspective. A perspective not overwhelmed by the chaotic deliria of the biodigital or a sightlessness that does not apprehend colonized spaces. Published in Italian in 1966, Rossi's book proposed a counter to the rampage of postwar modernist development. Rossi came to a position completely at odds with the zeitgeist of his times, a zeitgeist still informed by the verities of Congrès internationaux d'architecture moderne (CIAM), and its modernist revisionists. Rossi's positioned himself to be critical of the mega-structures and those postwar architects fascinated by technology (Tafuri, 1989, p. 135). As Tafuri (1989) was to argue, Rossi's poetics "placed him outside of contingent polemics" (p. 135). The sources of Rossi's position if we look for them are to be found in the social realism of the USSR, neo-realist cinema, a collaboration with Ernesto Rogers and Casabella, alongside debates in Italian architecture about the nature of urban history and the city – and Rossi's encounter with the urban science of French geography (Tafuri, 1989, p. 135).

Whatever else might be said about Rossi's views and corresponding poetics, we can at the least say that Rossi was against perceptual and psychological interpretations of the city; and perhaps it is these interpretations of the city that lead to delerium and sightlessness amongst architects. In his own projects, attention is paid to the formal, structural, and material urban phenomena of the city; there are no arcane discussions of scale. Rossi's position was not founded on a progressing history of events, linear narratives or processes. Rossi's disavowal of architectural narcissism – with its emphasis on the biography and personal creativity – resulted in an architectural language where "private and collective paths merged" (Tafuri, 1989, p. 135). Rossi's position vis-a-vis the city may, in the broader sense, point to how in current theory and practice, architects might disentangle and remake the encounter between climate and the city – in a way that might also suggest a kind of collectivity.

Architects, Sustainability and the Climate Emergency, 117–154

Published under exclusive licence by Emerald Publishing Limited
doi:10.1108/978-1-80382-291-420221005

Green Cities

Image Politics

The architectural construction of green cities is intimately linked to the channels of social media and concomitant image-politics of architectural discourse. Kevin DeLuca's *Image Politics: The New Rhetoric of Environmental Activism*, first published in 1999, serves as a useful guide to link together, the circulation of images in architectural discourse, to the politics of the good Anthropocene (DeLuca, 1999/2005). Hariman and Lucaites (2019) outline how DeLuca's book heralded approaches that began to examine visual media in relation to environmental activism (Hariman & Lucaites, 2019). Image politics is the enunciation and connection of "rhetorical criticism, media studies and politic action" (Hariman & Lucaites, 2019, p. 340). DeLuca's contribution was to posit how the image-politics of green environmentalism connected to the reimagining of the relationship between nature and civilization (2019, p. 340). In the field of architecture, a framework of image-politics, as it is denoted here, recognizes images as a critical means of expressing architectural polemics and associated rhetorics. Recognizing image-politics in architecture illustrates the various modes of visual representation and platforms through which green ideas are bought to life and disseminated in architecture. In focusing on image-politics in this chapter, it is assumed that in architecture, this politics encompasses both image, text, and the spatial manifestations of particular projects and their polemics.

This chapter focuses on the mediated space of digital platforms in order to understand the image-politics of architects and the projects that constitute the green city. To what extent do the rhetoric of architects, especially through social media, engage with the politics of the climate emergency? As Hariman and Lucaites' (2019) argue, an examination of image-politics "emphasizes the rhetorical effectivity of social movement rather than the organisation of social movements" and in turn this approach "shifts attention from the rhetoricity of words to a culture of images" with the result of recasting "the public sphere in terms of the public screen" (p. 341).

In analyzing the image-politics of architectural projects and how sustainable designs might constitute a green city, a number of principles should be borne in mind. First, images are presumed to be a primary means of communicative action in architecture's own political ecology. Although visual imagery is acknowledged in the rhetorical tradition, it is often located within a textual frame, subordinated to the power of the word, whether spoken or written: but no longer. In communicating to various publics, architectural ideas circulate through media channels – in this chapter digital magazines – in ways that present and signify environmental politics. This production of image events is now the primary – but not the only performance – for architects to disseminate ideas through social media. While environmental activism is the performance of particular events "within a mediated space defined by dramatic staging," this mediated space can now take place across social media (Hariman & Lucaites, 2019, p. 341). It is the digital magazine on which are projected the "public screen" of architecture. This screen can be regarded to: "explicitly positioned as a means of accounting for complex interactions of specific images, media logics, and economies of dissemination" (Hariman & Lucaites, 2019, p. 342).

While architects do not often act as environmental and disruptive activists – and perhaps this is the problem – an expanded notion of performance indicates how architecture is a spatial and image practice of public address. Moreover, the image-politics – and greenwashing – of the green city aligns with the virtue signaling studied in corporate marketing studies. For example, Gray, Sütterlin, Siegrist, and Árvai (2020) have studied company reputation and branding after environmental transgressions. They argue that consumers will cede to a firm a social licence if they feel their sustainability credentials are legitimate (Gray et al., 2020). However, for Gray et al. (2020), the context of corporate sustainability and the concept of social license is ambiguous and ill-defined; and results from a combination of different factors:

> …public or consumer trust in a company and its leadership, stakeholders' judgments about procedural fairness, the level of transparency of a company's practices, a company's compliance with social norms and expectations, etc. (p. 2)

Like corporations, architectural firms will attempt to build social license through various sustainability behaviors. On the one hand, these behaviors are established through evidence, such as a rating system, and are legitimate and earnest. However, at the other extreme is the idea of "greenwashing," in this scenario:

> Firms may, on the one hand, highlight symbolically significant sustainability-focused *programs* in order to deflect attention from a firm's environmentally unfriendly or less wholesome activities. (p. 2)

In the context of architectural firms the above deflection of attention may be evident but unintentional. It is perhaps easier to see how in their sustainable projects architects:

> …may selectively highlight specific, carefully selected sustainability *attributes* – e.g., certain behaviours, products, services, or even the corporate ethos (regardless of whether it's authentic or fabricated) – without drawing attention to potentially more important and relevant attributes or externalities. (p. 2)

Wallace et al. (2020) report on a quantitative study employing structural equation modeling to examine conspicuous virtue signaling (CVS) on Facebook. These authors study CVS in the reporting of donations to charities via Facebook profiles (Wallace et al., 2020). Their study, while focused on individuals and real vs counterfeit brands – rather than sustainable projects – suggests a parallel problematic to architects. This problematic is how the sustainability intentions of architects are projected onto different social media platforms. This study points to a number of questions for architects: When architects post their sustainable credentials on Instagram or Facebook, are they virtue signaling or building an extended identity to enhance their standing? Does this signaling accord with an architect's sustainable offline practices? Moreover, how do these image practices create an idealized green identity that gives the impression that an architect is achieving sustainability goals? Wallace et al. (2020) concluded that, amongst other things for individuals at least, virtue signaling was "a mechanism to bolster their self-concept enhancing their self-esteem" (p. 587). However, they also identified the paradox of virtue signaling as "a means of identity construction" and yet in the real offline world individuals – and dare we say architects – may have no intention nor the technical capability to achieve actual sustainable outcomes.

Dezeen's Sustainability Longlist

The green rhetoric and image-politics of architects point to a spectrum of behaviors that can be described as greenwashing through conspicuous virtue signaling in social media. In Dezeen, the online digital magazine, before Earth Day in 2021, an article entitled "10 ways in which architecture is addressing climate change" was published (Crook, 2021). The article described the different ways architects "can create a healthier planet" (Crook, 2021). The article highlighted the various trajectories architects have pursued regarding a so-called healthy Earth. These so-called top 10 ways include: building with timber, carbon neutrality, rewilding, Passivhaus, reversible design, nonextractive architecture, biomimicry, regenerative architecture, retrofitting and, last but perhaps not least, establishing climate action groups. The article points to the digital journalism tactics of, listing, the short summary, and boosting of particular projects or approaches. Green boosterism has increasingly been linked in cities research to the relationship between neoliberal practise and "pro-environmental branding strategies" (Garcia-Lamarca et al., 2021).

The Dezeen website describes itself as the most "popular an influential architecture interiors and design magazine" (Dezeen, n.d.). Launched in 2006, Dezeen was acquired by JP/Politiken Media Group, a Danish media group and Denmark's leading media company. Prior to the acquisition this group had revenues in 2020 of 2.998M DKK (0.475M$US) (JP/Politiken Media Group, n.d). The group is owned by two private trusts "committed to social responsibility and believes that journalism is vital to democracy" (Ravenscroft, 2021). In the context of global media, Dezeen's revenues are modest. Dezeen is part of the Federation of Small Businesses (FSBs) and the Good Business Charter, a Charity registered in England and Wales with both union and industry representation on its board (Dezeen, n.d). Dezeen's founder Marcus Fairs is described as a digital journalist. Moreover, Dezeen also operates a design jobs board, an events guide, a virtual showroom for brands and products, and an annual awards program. Dezeen is ranked as one of the most visited websites in the world. Dezeen's longlist for its 2021 sustainable building awards indicates the way that information about architectural projects and green cities circulates in social media. The Dezeen sustainable longlist speaks to prevailing sustainability preoccupations in architectural discourse (Dezeen Awards, 2021a).

The circulation of green architecture through social media and online platforms like Deezen conceals the complexity of carbon mitigation. In the past, an architectural firm's reputation was based on its ability to deliver a project, create relationships with its clients, or develop projects with significant symbolic capital. Weld Coxe, in the mid-1980s, argued that practices move from being practice-based firms focused on design and then over time transform into business-based practices with a focus on service and delivery (Coxe et al., 1987). This framework does not account for digital transformation, new procurement methods, social media, and how branding shapes the perception of architects by various clients. Nor does the framework suggests how architects might develop business models that directly address sustainable practice across all of these dimensions (Coxe et al., 1987).

Architects are now judged on both their ability to deliver sustainable designs, and how they are perceived as sustainable designers – perhaps more so. As suggested in the preceding section, architects may also be prone to conspicuous virtual signaling and virtual consumption:

> …if that conspicuous consumption is about "doing good," to what extent does this affect self-esteem and predict offline prosocial (donation to the charity) or even unethical (counterfeit purchase) behaviour intentions?
>
> (Wallace et al., 2020, p. 578)

All of the projects in the Dezeen 2021 sustainable longlist present a range of sustainability claims. Yet these seemingly concrete claims do not often point to the ambiguities inherent to the sustainability of each project (Dezeen Awards, 2021a). Typical of examples on the longlist are those projects with certification claims, for example the Olympic House, the IOC's new headquarters in Lausanne – designed by Copenhagen's architect 3XN (Ravenscroft, 2019). Dezeen states the building "optimises energy and water use and enhances employee well-being" – we might ask how it achieves this latter attribute (Dezeen Awards, 2021b). The building has achieved high ratings with LEED Platinum, reportedly with the highest score ever given, Swiss Sustainable Building Standard (SNBS), and the related Swiss Minergie certification system. As Dezeen proclaims, these are "three of the most rigorous sustainable building certifications" (Dezeen Awards, 2021b).

And yet, the history, development, and ongoing research into the efficacy of different rating tools, as discussed in detail in the next section, are a quagmire of fragmented claims and counterclaims (Ade & Rehm, 2020). Certainly, each project attached to a particular rating system may indicate an architectural firm's cachet in regard to its sustainability efforts. However, all of these different tools are the result of different histories of development and use. Across the globe different rating tools, regulations and institutional structures operate. Ade and Rehm (2020) in a historical survey of these tools and their development include: BREAM, LEED – "has certified less than 1% of US buildings" – Greenstar, and, in Australia, NABERS (Ade & Rehm, 2020, p. 13; Lorch, 2017). Recognizing how these tools are governed by Green Building Councils (GBCs), Ade and Rehm (2020) note that:

> This history indicates that from inception the development of green building rating tools has been predicated on the needs of a very specific grouping of organisations (the members of GBCs) rather than the wider needs of the planet. (p. 13)

As well, the limited take up of the tools has been hampered by property developers who appear only interested in the branding that certification from a particular tool confers on a building. They state that:

> A performance rating that is issued only after one year of operation is of little use to a developer and this may be why none of the original rating tools were designed as actual measured performance rating tools.
>
> (Ade & Rehm, 2020, p. 13)

Another rating tool or system is one promoted by the International Living Future Institute (ILFI) – and surprisingly not covered by Ade and Rehm (2020). This system is represented by two other buildings in the Dezeen longlist. First, the architects of the Catalyst Building, a mass timber CLT building in Spokane Washington, claim to be pursuing "Zero Energy and Zero Carbon" certification from ILFI (Dezeen Awards, 2021c). Second, the Kendeda Building for Innovative Sustainable Design at Georgia Tech is another building seeking ILFI certification – on the longlist and then shortlisted by Dezeen – from the International Living Future Institute (Dezeen Awards, 2021). For this building it is claimed to have "full" Living Building Certification (LBC) after establishing "its net positive energy and water performance during its year-long occupancy period." LBC is a performance-based system sponsored by the ILFI (International Living Future Institute, n.d). ILFI claims that the LBC is superior to LEED because it is a flexible, rigorous, and holistic system of certification. ILFI claims that projects that reach LBC certification "can claim to be the greenest anywhere." With the objective of transforming "every single act of design and construction," the system was the 2012 winner of the Buckminster Fuller Prize (International Living Future Institute, n.d). The system is a performance-based methodology that allows it to be directly employed, adapted, and crafted by design teams in order to match certification with a particular project. Moreover, the methodology is constantly updated as a result of feedback from project teams; consequently, LBC has undergone a number of iterations and is now at version 4.0. The challenge is structured through seven overarching "petals" which organize seven different "imperatives." The petals are place, water, energy, health and happiness materials, equity and beauty (International Living Future Institute, n.d).

A number of buildings on the list also claim to have been designed as a result of Passive House (German: PassivHaus) principles. One of these buildings, the Woodside Building for Melbourne's Monash University, by Grimshaw Architects, is described as the largest Passive House building in the world (Dezeen Awards, 2021a). Also on the list is the first major Passive House buildings in Cambridge for King's College, Cambridge, this comprises graduate student accommodation for 59 students and the Haycroft Gardens "energy-efficient multi-generational family" home by Sarah Wigglesworth Architects (Dezeen Awards, 2021a). The Woodside building has been the focus of controversy because the building was completed by Monash University as a result of its partnership with Woodside Energy, a large Australian fossil fuel producer. The research partnership includes Carbon Capture and Storage projects. In October 2021, Extinction Rebellion protestors glued themselves to the glass walls of Woodsides HQ in Perth. This protest was to declaim Woodside's Scarborough Gas Project (Tilly, 2021). The gas field is calculated as contributing 1.69 billion tonnes of CO2 over its 30-year life and potentially damaging rock art of indigenous global significance in the area (Clark & Tilly, 2021).

Meeting Passive House criteria in a particular jurisdiction might signal that a building or building project is green. However as indicated above, citing Passive House (PassiveHaus) in social media may obscure a more complex sustainability or political narrative underlying a project. In addition, Niskanen and Rohracher (2020), in an examination of two Swedish Passive House examples, assert how the uptake of this standard may change between regions (Niskanen & Rohracher, 2020). Reports of certification instances in media such as Dezeen may conceal the factors related to sustainability and state-based innovation systems. In their Swedish study, Niskanen and Rohracher (2020) outline where the concealment of complexity, and not simply a quantitative complexity, is likely to take place. This concealment is the result of the tendency to focus on discrete, singular, and atomistic sustainability attributes. Countering this, they outline the importance of organizational issues:

> A mere focus on actors and technologies directly involved in the planning and construction of passive houses risks failing to understand important aspects of their role in organisational change.
>
> (Niskanen & Rohracher, 2020, p. 3)

Second, how a particular standard, or certification system, is communicated is also of concern:

> For passive houses to be perceived as successful innovations, they need to be communicated as an improvement to both their

> different users as well as to decision-makers, and board managers or other relevant audiences.
>
> (Niskanen & Rohracher, 2020, p. 3)

With the above issues in mind, Niskanen and Rohracher (2020) conclude that:

> Institutional, social, and material dimensions in the wider context of passive houses might facilitate or hinder their uptake: for example, local policies, networks, and practices might guide resources towards alternative energy efficiency solutions or alternative sectors. (p. 3)

Given the issues that surround how architects communicate their own value to various public's, the above findings suggest the pitfalls of sustainability headlines (Samuel, 2018; Raisbeck, 2019).

A number of the buildings on the longlist claim sustainability and virtuous capital through new methods of innovative construction. These include: the Braunstein Taphouse, a building in Koege Denmark, a project able to be disassembled to enable its building components to be reused (Dezeen Awards, 2021a). The Sands End Arts and Community Center, designed by Mae community center, in the Borough of Hammersmith and Fulham, utilizes CLT timber (Dezeen Awards, 2021a). An energy self-sufficient prototype, designed by students at TU Delft called MOR – an acronym for Modular Office Renovation – is intended to transform existing office space to "net-positive" space; the resultant 50m^2 dwelling incorporates climate design strategies that align with indoor comfort and "low energy demand" (Dezeen Awards, 2021a). Finally, there is also the ciAsa Aqua Bad Cortina project by Pedevilla Architects, a project in the South Tyrol (Dezeen Awards, 2021a). As Schlogel et al. (2020) note, natural hazards in the South Tyrol, such as "debris and mud flows, landslides, avalanches, rock falls, and (flash) floods," are linked to climate change. This latter project is constructed from spruce wood that became available after a severe storm. While the façade employs Larch shingles hand split by local craftsmen (Dezeen Awards, 2021a).

Craft traditions and associated sustainability innovations are also highlighted in the longlist; craft traditions, perhaps because of their connotations of past socio-natures, may more easily appear to signify sustainability. A number of these projects appear to highlight issues of local craft in local and seemingly vernacular contexts. This includes the Arc, designed by IBUKU, which is constructed from a gridshell bamboo structure at the Green School in Bali, Indonesia. Designed by Atlas Studio, The Well House is a "boutique hotel" that aims to easily fit into the context of the "austere lifestyle" of a 600-year-old Dong village in Guizhou Province China (Dezeen Awards, 2021a). Another project in the PRC longlisted by Deezeen is the "The House of Dreams," which is a project centered on the cave area in Zhoushan. In this project, the caves were the childhood home of villagers, and these are now incorporated into the project. In addition, the new buildings in the project are made from "recycled construction waste materials" and constructed by the villagers themselves (Dezeen Awards, 2021a).

One notable project on the list that points to the problems of virtuous signaling and the ambiguity of sustainability claims and sustainable image-politics is the Integral project in China. This project was designed by the Hong Kong firm Ronald Lu and Partners; as a shortlisted Dezeen project in Guilin, China, this project is claimed to be "fusing advanced technology with conservation elements and humanistic architecture" (Dezeen Awards, 2021d). A project, entitled Silk Pavilion II, designed by no less than the American digital architect and media artist by Neri Oxman, was unveiled at the Learning Center at Integral. As Ronald Lu states in Dezeen:

> Integral is a place where people, technology, manufacturing and nature co-exist in harmony. Balancing sustainability with the principles of a Chinese garden and cutting-edge industrial practices, the project consists of a state-of-the-art production facility, an exhibition zone showcasing the technological evolution of the textile industry, a learning centre, picturesque staff facilities, a central wetland area and a botanical garden where natural dye crops grow.
>
> (Dezeen Awards, 2021d)

As described in Design Anthology, the "luxury interiors, design, architecture and urban living magazine":

> Integral is a unique project because it embodies sustainable principles. To reduce the environmental impact of the project, it combines advanced textile industry technology with an eco-tourism experience.
>
> (Design Anthology, n.d)

The project was designed for Marjorie Yang, a "second_generation" industrialist and chairperson of textile manufacturing company Esquel. Esquel supplies garments to both Tommy Hilfiger and Nike (Business & Human Rights Resource Centre, n.d). The Esquel textile company is currently pursuing legal action against the US Department of Commerce to remove the company's "erroneous" inclusion on the US entity list. Up until 2020, Esquel was a member of the Fair Labor Association, prior to allegations that it was using forced labor in its cotton mills. In 2014, Esquel set up the Xinjiang Research and Development Center. This center was set up two years prior to the instigation of the Xinjiang internment camps for the Uighur people (Xiuzhong et al., 2020). In 2018, Esquel received accreditation for one of the cotton species it had developed from the Xinjiang Uygur Autonomous Region (Esquel Group, n.d). However, in September 2020, in response to a Wall Street Journal article in May 2019, Esquel issued a statement, citing its long history in Xianjing, about its operations in the Xianjing province. The company denied using Uighur workers in unethical employment practices

(Esquel Group, 2020). The company is currently pursuing a lawsuit against the US Department of Commerce to remove the erroneous inclusion of Changji Esquel Textile Co. Ltd. (CJE) on the US Entity List.

Green Cities in China

The above is not to excoriate the work of Chinese architects or projects. The necessary point in examining the politics of the Integral project is to reveal how supply chains are interconnected with the circulating image-politics of virtuous sustainability. For numerous sustainable architects, urban designers and landscape urbanists, Chinese cities are seen as an open field of opportunity where sustainable tactics can be enacted. The journal *Architectural Design* has also paid attention to the contemporary architecture of the People's Republic of China (PRC). In a project published in *Architectural Design* in 2017, Richard Garber sees the opportunities in engaging with China and seeing it as a "great urbanistic opportunity," he cites the example Zhangdu Lake Farm ecological village, a project that couples together urban development and existing land-use practices through processes of community participation, digital design technologies, and "green thinking" (Garber, 2017). In 2018, Architectural Design published a special issue on Chinese experimental architecture entitled China Homegrown: Chinese Experimental Architecture Reborn (Lu & Dou, 2018). The Cambridge University educated editors of the issue sought to position themselves in relation to design research, and a number of articles in this issue address notions of sustainability. For example, the editors published the work of Xiao Fu and Wei You from Nanjing University, who argued that local Chinese cultural differences must be accounted for when considering how to connect together local vernaculars, sustainability, and spatial tectonics (Fu & You, 2018). As might be expected, Architectural Design like Dezeen continues to be a channel for the various sustainability proclamations of PRC located projects. Vicente Guallart, former Chief Architect of Barcelona, in the March 2020 AD issue "Digital Cities to Biocities: Harnessing the Power of the Digital Revolution to Reinvent the Urban Ecology Model," describes his practices winning competition – design for a "green" district, Xianmihu, in Shenzhen, China. He claims that the project represents a new "natural/urban digital synthesis" (Guallart, 2020).

The green cities and projects produced by architects, and fueled by the laziness of good intentions, represent the image-politics of a business as usual Anthropocene. The homogeneous and unitary images of this greenness, which platforms like Dezeen depict, often belie a milieux of fragmented and ambiguous, technological, organizational, and unacknowledged political conflicts. And then, as discussed below, so-called green materials are deployed to promote sustainability credentials – we might designate this the city of green materials; through which these materials, like much else in architecture, are used as signifiers, as virtuous signals, by architects; only serving to reinforce their own status, as technological innovators, as climate fixers and polemicists and theorists; and often with the best of intentions. However, in this mileux of green materials, the noise generated by these performances appears to obscure the continuing production of carbon emissions through entrenched material and product systems.

Embodied Emissions and Carbon Ambiguity

The debates around embodied carbon in construction materials are inextricably linked to Life Cycle Assessments (LCAs) and Life Cycle Inventory (LCI) Analysis. LCAs measure the environmental impacts linked to products and product systems over their entire life cycle (Klöpffer, 2014). LCA methods are described in the international standard ISO 14040 of 2006 standard and defined as:

> Life Cycle Assessment (LCA) is a science-based, comparative analysis and assessment of the environmental impacts of product systems. It is distinguished from other environmental assessment methods by two constitutive and unique features: the analysis from 'cradle-to-grave' and the 'functional unit.'
>
> (Klöpffer, 2014, p. 2)

The ISO 14040 and 14044 standards are the framework for employing LCAs to interrogate various product systems (ISO, 2006). The first LCA standard was released in 1997, revised three times (14040 14041, 14042). It was then updated in 2006 with the combination of 14042 and 14043 to become ISO 14040 and 14044 (Pryshlakivsky & Searcy 2013, p. 115). Crawford, Stephan, and Schmidt (2018) suggest that the ISO 14040 standard gave the impetus for LCI, and hence embodied carbon research with the work of researchers such as Treloar (1997), Joshi (1999), Suh and Huppes (2002), and Crawford, Bontinck, Stephan, Wiedmann, and Yu (2018, Fig. 1, p. 1276). LCI methods are the second step, within the LCA standard, that includes the following steps:

> (1) Goal Definition and Scoping (2) Inventory Analysis (Materials and Energy Acquisition; Manufacturing; Use; Waste management) (3) Impact Assessment (Ecological Health; Human Health; Resource depletion) (4) Improvement Assessment.
>
> (ISO, 2006, p. 2)

Hence, LCI analysis is a central component of all LCA studies. LCI methodologies and information are extremely important when considering how architects make design decisions regarding material choices in relation to embodied carbon and its environmental impact. LCI analysis is concerned with a description of specific inputs and outputs for a given product system – in other words, a list or table of inputs and outputs (Crawford et al., 2018, p. 1274). An LCI approach using an LCI database helps to articulate the specific inputs and outputs related to the material components of a particular building. In a review of LCI methods, Crawford et al. (2018) note the three approaches to developing an LCI. These can be summarized as: a bottom-up approach where the product system is analyzed as a series of processes throughout its lifecycle; a top-down macro-economic environmentally extended input-output analysis; and finally what they describe as "hybrid analysis" which involves combining, and presumably judging the interface between, the top-down and the bottom-up approaches. In short, hybrid LCI models describe a product system through both "specific production processes or entire economic systems" (Crawford et al., 2018, p. 1274).

Increasingly embodied carbon has become the focus of LCI methods. As Röck et al. (2020) note, some of the previous work around embodied carbon is now redundant. In previous studies of buildings, embodied carbon was seen to comprise only 10% of operational emissions. Consequently:

> ...the embodied contribution to life cycle energy was within the uncertainty range of the energy demand forecast for building use and thus not considered relevant. However, this situation has changed dramatically. In recent years, several studies have demonstrated the growing importance embodied impacts, both relative to their contribution to life cycle-based performance and in absolute terms. (p. 3)

Minunno et al. (2021) in a systematic literature review, and meta-analysis of embodied energy carbon in buildings, seek to overcome what they described as the failure of previous LCA studies to provide "a consistent picture" of embodied energy and carbon in different construction materials (p. 1). The aim of their study was to provide direct and practical advice to LCA practitioners. The review included 73 articles on construction materials, 100 articles on buildings, and 48 articles on LCA strategies and insights (Minunno et al., 2021, p. 3). By extracting data from the literature, a picture of both embodied energy and carbon was developed for eight construction materials: concrete, reinforcement bars, steel, timber, bricks, tiles, insulation, and plaster (Minunno et al., 2021, p. 5). One notable aspect of the research was to graph the coefficients of embodied energy and carbon in different materials between different countries, and this indicated a wide degree of variation (Minunno et al., 2021, Fig. 5, p. 7). This analysis also included an analysis of the embodied energy across concrete production in different countries (Minunno et al., 2021, Fig. 6, p. 8). As a result of this survey and analysis, the authors suggested that the LCA assessment method of four steps would benefit from the addition of two further steps. In relation to timber and concrete, amongst other things, these authors found that there were real benefits in replacing concrete with timber:

> Using timber instead of concrete, whenever feasible, allows a decrease in buildings' environmental impact up to 43% in terms of embodied energy and 68% in terms of embodied carbon. (p. 11)

However, this finding is not without its nuances in regard to timber – when compared to concrete – when the difference between the embodied energy in timber production and timbers embodied carbon:

> ...although trees capture carbon from their environment, the industrial processes involved in harvesting, drying, sawing, and transporting timber are energy-intensive. The gap between the embodied energy of timber structures and their embodied carbon has been assessed and can be explained by timber absorbing CO_2 from the atmosphere, contributing to a low amount of embodied energy. (p. 5)

A number of databases have been developed in recent years that indicate, for architects and other built environment experts, the embodied carbon in different materials. For example, Mohebbi et al. (2021) in a study focused on the Inventory of Carbon and Energy (ICE) database, and the GHG conversion factor data, published in 2020 by the UK Department of Energy and Climate Change, it was found that: "EC [embodied carbon] calculations can vary drastically based on the chosen database" (p. 1). These authors argue that in built environment practice and projects, there has been an over emphasis on operational phases in particular energy use – perhaps because of the focus of this in IPCC reports. However, as they emphasize, emissions in construction and associated GHG emissions "will enter the atmosphere earlier, and the negative impacts will be seen across an extended period" (p. 2). Mohebbi et al. (2021) conclude by bemoaning the lack of a national and consistent product manufacturer database:

> If an LCA is carried out as part of the early design process, the ECF [Embodied Carbon Factor] database would need to provide multiple alternatives. The consistent collection method within an ECF database would guarantee that the comparison of different designs would be viable.
>
> (Mohebbi et al., 2021, p. 21)

Some of these issues are overcome in Australia, where there has also been important research related to embodied carbon databases (Crawford et al., 2018). The EPiC database is an open access LCI inventory of embodied environmental flow coefficients (Crawford et al., 2021; Crawford, personal correspondence, 2021, October 4; Crawford, Stephan, & Prideaux, 2020). The designers of this database point to the complexities of measuring embodied carbon:

> Although various other LCIs and similar databases exist, they often suffer from inconsistencies in system boundaries, lack of data transparency and variations in temporal and geographical specificity.
>
> (Crawford et al., 2020, p. 2)

The EPiC database presents coefficients and data about three different types of environmental flows, energy, water, and greenhouse gas emissions. The EPiC database has a number of unique characteristics that are different from the United Kingdom's ICE database. It comprises a consistent method in calculating the 284 coefficients across materials as well as employing transparent and publicly available data for each material. Importantly, for each material in the database, the coefficients are based on a complete system boundary; thus, everything in the supply chain is covered. This is a more independent and transparent process than simply sourcing material from product manufacturers, who may not always have information across the supply chain for a product – maybe it goes without saying, but architects need to question carbon data form product manufacturers (Crawford et al., 2020).

EPiC was released in late 2019. Architects have a critical role to play in adopting these databases and using them to make early-stage design decisions. Large engineering consultancies which provide specialist environmental sustainability services have also begun to use these types of LCI databases; foreseeably the use of these databases will become a professional and, indeed, an ethical obligation. Moreover, market uptake of such databases in construction and product markets obviously requires market demand.

For many architects timber and timber product systems are perceived as containing less embodied carbon due to timber's association with organic processes of manufacture. Timber is thus less associated with carbon-intensive manufacturing processes as well as being associated with carbon sequestration as timber is grown, and hence the use of timber is thought to have less impact in comparison to nonbiological materials. However, confirming "definitive conclusions" in accounting for carbon in bio-based materials is not possible (Morris, Allen, & Hawkins, 2021, pp. 1–2). This is because there are problematic issues in regard to bio-genetic materials: different treatments in LCA calculations, a lack of consensus in LCA calculations, and issues regarding the timing and duration of emissions. These authors concluded that:

> …when considered on a cradle-to-gate basis (A1–A3) – as has often been done in past studies – glulam options have substantially lower GWPs than their steel counterpart when biogenic carbon storage is included.
>
> (Morris et al., 2021, p. 5)

They noted that any benefit of a bio-based product such as glulam (CLT) timber was subject to several variables:

> …the end-of-life scenario assumed for timber; the treatment of biogenic CO_2 and; the consideration of emission timing.
>
> (Morris et al., 2021, p. 9)

To reiterate, the problematic nature of embodied carbon accounting is a reality. Embodied carbon accounting is not well developed in architecture, nor more broadly understood in design and construction, arguably the take up of embodied carbon databases might have parallels in the slow industry and market take up of green rating systems. Richard Lorch (2017), in a *Building Research and Information* editorial, summarized the rating situation by identifying policy failures in a number of key areas. Lorch (2017) pointed to the minimal development of new green buildings through the low uptake of market-led rating tools such as LEED and BREAM. They see these tools as only having an impact in "niche markets" (p. 475). In relation to what he designates research and practitioner communities, and this presumably includes architects, Lorch (2017) applauds innovative projects Passivhaus and other regenerative design systems like LBC. However, they bemoan the lack of wide adoption and influence of these approaches because of ineffective policy and regulation drivers. As an antidote, Lorch (2017) calls for policy approaches that are simultaneously "top-down, bottom-up, and middle-out as well as being 'interlocking'"; not simply with a focus on regulation (p. 476). For architects, their own claims regarding embodied carbon often go uncontested; the complexities and ambiguities of measurement often remain unspoken or unexamined; at worse, ambiguity is glossed over in favor of green images. Identifying greenwash, and untangling these claims from their broader policy and political contexts, point to the failures in terms of embodied carbon accounting and GHG reduction in the design and construction sector.

The Materials Book

The Materials Book, published in 2020, exemplifies the way that green materials, embodied carbon, and sustainable construction processes have been thought about by architects (Ruby & Ruby, 2020). With an introduction by Marc Angélil and Cary Siress of ETH Zurich, the book is published by Ruby Press and edited by the press's owners Ikea and Andreas Ruby (Angélil & Siress, 2020). The book is a comprehensive compendium – Whole Earth Catalog style – that includes material developed by LafargeHolcim Foundation for Sustainable Construction and the contributions to the sixth international LafargeHolcim forum for sustainable construction. This forum took place at the American University of Cairo, April 4–6, 2019. LafargeHolcim is a Swiss company whose largest revenues are derived from cement, clinker, concrete, lime, and aggregates. The Materials Book points to how notions of materials – and materiality however it is defined – have been theorized and approached by architects.

In their introduction to The Materials Book, Angélil and Siress (2020) proclaim that the "material intensity of mainstream building is unsustainable" (p. 11). The construction industry's destruction of the environment is posed as a binary: the difference between "dematerialization" versus "rematerialization" (pp. 13–14). In other words, the reduction of physical materials in construction – "doing more and with less and less until eventually you can do everything with nothing" versus the ideas of so-called cradle to cradle manufacturing concepts (Braungart & McDonough, 2009). This latter approach is described as:

> ...a twofold pathway toward reframing the entire sector by promoting both building materials that have little impact on the

> environment and the repeated reuse of materials after their conventional use cycle,
>
> (Ruby & Ruby, 2020, p. 14)

The idea of realizing a "sustainable materialization" in the built environment (Ruby & Ruby, 2020, p. 23) is presented in the form of a polemic of 22 propositions for "Re-materializing Construction." The 22 propositions are a smattering of different aphorisms. For example, "How Much Does Your Building Weigh Mr Foster," "Do Not Forget CO2," "Produce Leaner," "and 'If less is More, Maybe Nothing is Everything.'" This section is edited by Sarah Nichols, an Assistant Professor at Rice University, with the aim to "dematerialize construction by rethinking the building-material cycle from extraction to processing, design, transport, installation, maintenance, and removal" (Ruby & Ruby, 2020, pp. 25–56).

Overall, the book is a comprehensive snapshot of the plethora of viewpoints and approaches architects have toward construction materials and climate. Following the 22 propositions, there are 32 other contributions, including the keynote architectural speakers from the 2019 forum: Christine Binswanger, Lord Norman Foster, and Lalai Iskander. Like the 22 propositions, this compendium is then organized around a number of sub-themes with catchy titles. For example, "Changing Paradigms: For a World Not Yet Built," "Shifting the Flows, Pulling the Strings: Stocks, Flows, and Their Dynamics," and "Catch-22: Material Needs versus Material Impact." Some of the contributions are little more than vague polemics and calls to action. Mushrooms and mycelium are discussed in two contributions. Ideas of circular economy and industrial ecology appear in a number of others. Importantly, there are also contributions from architects and practitioners working in the Global South. This includes Eva Pfannes's analysis of water collection and distribution systems in Chennai, India, described in two-dimensional flow diagrams (Pfannes, 2020). Moreover, the diagrams of Alla and Khalil (2020) are also cogent as they describe the material flows of waste in Cairo using the kilogram as the basic unit of measurement (Alla & Khalil, 2020).

Much of the rhetoric in the book likens cities to ecosystems or living organisms. And there is also a fascination with the weight of materials and achieving carbon mitigation through product systems that are argued as lesser or leaner (Proposition 29: Let scarcity be your inspiration). Many of the contributions appear formulaic and follow a familiar pattern: beginning with a problem statement of the earth's plight, which is then solved by a particular mindset, technology, or process. Presumably, LafargeHolicim would be happy with all of these positive and varied contributions to the carbon-emitting dilemmas of their core business. In the noise generated by the above, there is scant discussion regarding the ambiguities of carbon emissions metrics nor much about LCA and LCIs. Everything in the book is pitched as a solution. From this viewpoint, it is easy to conclude that this noise, emanating from the forum, creates a haze that disguises the business as usual approach of product systems embedded in cycles of capitalist accumulation.

Interestingly, Sarah Nichols (editor of the tome's 22 propositions) had previously explored the power relationships in the production of cement and concrete through the diagrams of the work of Georges Baehler, a European communist and Anti-Fascist hydroelectric engineer (Nichols, 2018). Nichol's concluding point in that article regarding Baehler highlights the political ironies of the forum:

> The business of cement, and of building materials at large, extracts, produces, and normalises; it seeks energy sources, arranges logistical networks, and commissions infrastructure. That is, through the immaterial it produces and continually adapts an apparatus that perpetuates material demand. (p. 150)

In 2021, a group of climate activists took direct action against LafargeHolcim in order to save Mormont Hill, north of Lausanne, a habitat for wild orchids. The group alleged that the company was going to destroy in the process of mining limestone for cement production. The activists occupied the hill in March 2021 in order to stop the expansion of the cement factory and the destruction of the hillside. This was an early morning nonviolent action. The activists argue that the Hill has significant archaeological, biodiversity, and landscape values: They proclaim:

> Holcim Switzerland is seeking to obtain new permits as soon as possible to continue the unbridled exploitation of the Mormont hill. If this is not stopped by 2029, the entire Birette plateau will have disappeared.
>
> (Zad de la Colline, n.d)

In October 2021, the group was seeking to raise funds to support the legal costs incurred as a result of fines imposed as part of their occupation campaign (Zad de la Colline, October 25, 2020).

Certainly, climate issues related to cement production and concrete construction can be seen as a site of conflict. For example, following the publication of the IPCC AR6 Working Group I report in August 2021, Dezeen put into circulation information and ideas about concrete construction and carbonate emissions. In the article, Richard Leese, a lobbyist at the Mineral Products Association, which represents the United Kingdom's cement industry, is reported as saying "The key chapter in [the IPCC report] recognized concrete carbonation for the first time." The article in Dezeen that was published days after the AR6 report suggested in its title the claim that "Concrete construction 'offsets around one half' of emissions caused by cement industry says IPCC" (Fairs, 2021). The article then went on to say that:

> Around half of the carbonate emissions from cement production are reabsorbed by the material when used in buildings and infrastructure, according to the latest IPCC climate report.
>
> (Fairs, 2021)

Arguing that:

> The "cement carbonation sink" absorbs an estimated 200 million tonnes of carbon every year, according to an overlooked section of the report published earlier this month ahead of the COP26 climate conference.
>
> (Fairs, 2021)

In other words:

> "The uptake of CO2 in cement infrastructure (carbonation) offsets about one half of the carbonate emissions from current cement production." However, carbonate emissions only represent between 40 and 60% of total emissions from cement production so carbonation will reabsorb around a quarter of the emitted carbon.
>
> (Fairs, 2021)

Immediate comments in response to the article from members of the UK ACAN network pointed to the obvious misinformation promoted by the article. Claiming the original headline using the word "half" to be an example of "greenwashing" and climate hoaxing. In the comments section, it was pointed out that:

> Carbonate emissions refer to the emissions associated with driving off CO2 during calcination. This is only a portion of the *total life cycle emissions* of carbon dioxide from the cement industry (and concrete industry for that matter). This article is very deceiving and should be retracted or corrected significantly.
>
> (Mikaela, 2021)

In response to the comments, author and founder of Dezeen Marcus Fairs changed the heading of the article and issued the following statement:

> We are certainly not funded by the concrete (or cement) industry but it seems the potential for concrete and cement to sequester significant amounts of atmospheric CO_2 has been widely overlooked. However, according to the IPCC report, it only absorbs half of the carbonates

> emitted during cement production in the first place (and none of the fossil emissions as far as I understand it) so the industry is still a major contributor to global warming.
>
> (Marcus/Dezeen Fairs, 2021)

As noted in one comment on Dezeen, the offsetting of carbon emissions as a result of producing more concrete as a "carbon sink" (via carbonation) was obviously self-defeating.

The empirical uncertainties of embodied carbon are rarely stated in architectural discourse. Apart from the work of ACAN, discussed in Chapter 6, the actual material ambiguities of embodied carbon, outside of the image-politics of architects – timber is good vs steel is bad – are rarely discussed. To remedy this, architects must turn to three areas of knowledge. An understanding of the historical science of embodied carbon metrics, the need to adopt practical approaches to embodied carbon measurement – including any inherent uncertainties – as well as aggressively to advocate, for the widespread adoption of these measurements across different policy settings and by urgent necessity through architectural education. For if all else fails, and such things are too hard, architects can return to the easy ebbs and flows of virtuous image-politics.

Biophilic Cities

Floating amongst the flotsam and jetsam of incremental innovations and greenwashed media circuits of capital, is the idea of biophilia. This is the idea that humans are inextricably linked to the natural world and this linkage has a genetic basis in humans. Biophilic design and placemaking has been seen as the vehicle by some architects to mitigate and adapt to global warming. Biophilia has attracted the interest of numerous architects in the areas of urban design, healthcare design, workplaces, and a range of housing typologies. For example, Richard Mazuch, a healthcare specialist at the global IBI design group, writing in *Architectural Design* in 2017, claimed that research supports the idea that positive health outcomes can be gained from biophilic design (Kellert et al., 2011; Mazuch, 2017). He states:

> Recent research supports measurable, positive outcomes of biophilic design on health. Biophilic design can clearly improve well-being and wellness and can expedite healing within a world of increasing populations and urbanisation and increasingly hermetically sealed living spaces.
>
> (Mazuch, 2017, p. 44)

The biologist Edwin Wilson is regarded as the original thinker, for want of better words, of the term biophilia. Wilson's initial conceptions of biophilia serve as an origin to understand how the term has been translated and utilized by architects. Writing in the Prologue to his book, Biophilia Wilson was to describe an epiphany.

> On March 12, 1961, I stood in the Arawak village of Bernhardsdorp and looked south across the white-sand coastal forest of Surinam. For reasons that were to take me twenty years to understand, that moment was fixed with uncommon urgency in my memory.

Wilson's book is part memoir, travelogue, reflection, and dream journal:

> The emotions I felt were to grow more poignant at each remembrance, and in the end, they changed into rational conjectures about matters that had only a distant bearing on the original event. The object of the reflection can be summarised by a single word, biophilia, which I will be so bold as to define as the innate tendency to focus on life and lifelike processes.
>
> (Wilson, 1986, p. 1)

In reflecting on biodiversity, Wilson's central hypothesis is that:

> I have suggested that the urge to affiliate with other forms of life is to some degree innate, hence deserves to be called biophilia.
>
> (1986, p. 85)

Going on to say:

> The biophilic tendency is nevertheless so clearly evinced in daily life and widely distributed as to deserve serious attention. It unfolds in the predictable fantasies and responses of individuals from early childhood onward. (1986, p. 85)

Wilson's hypothesis is further developed in his 1993 book (republished in 2013), bringing together a number of viewpoints and perspectives of biophilia:

> Biophilia, if it exists, and I believe it exists, is the innately emotional affiliation of human beings to other living organisms. Innate means hereditary and hence part of ultimate human nature.
>
> (Kellert & Wilson, 1993/2013, p. 31)

With the above in mind, it can be posited that design strategies based on biophilia theory have been seen as a way for architects to plug the gaps – and not simply conceal – sustainability transitions. For example, Ramzy (2015) argues that the obvious deficiencies of sustainable design can be augmented with authentic ideas of biophilia. Ramzy is aware of the dangers of employing both biomimetic and biophilic design; noting that architects must "go beyond simply mimicking Nature" (Ramzy, 2015, p. 248). As Ramzy correctly points out biomimetic design – as compared to biophilic design – may not necessarily fuse itself with nature in the same way that proponents of biomimetic and biodigital design claim. Ramzy's project is to "supplement biomimetic applications with biophilic qualities" (Ramzy, 2015, p. 249). In pursuing this agenda, they seek to locate the biophilic impulse in a broader canon and history of architecture. To reclaim biophilia in the canon, they argue that contemporary architecture must recognize: the iconographic and compositional parallels of biophilic practices, typologies, and configurations in historic architecture. Ramzy is a scholar of medieval and Coptic architecture, and their project is an effort to embed biophilic design into contemporary architecture. This is done through a matching process. For example: matching the Quarter Cacti building with Mashrabiya screens, matching Mark Sarkisian's project for China World Trade Center replete with Gothic cathedrals, and matching the Eastgate Center in Harare, Zimbabwe, designed by the architect Mick Pearce. This latter project employed a climate control system inspired by termite mounds. In advancing this argument, Ramzy seeks to counter expressions of biophilic design that have been reduced to simple functional and additive elements. Rightly, sensing in this prevailing approach an architecture "reduced to the state of 'construction,' rather than 'design'" (Ramzy, 2015, p. 55).

Demers and Potvin (2017), like Ramzy (2015), argue that biophilia should not be reduced to a few simple precepts. These authors employ a design research methodology to proclaim that biophilia is "more than the mere human to nature transactions." They hint, somewhat opaquely, that biophilia and the development of an affinity to nature can "emerge from the actual experience of natural forces through the creative process" (2017, p. 326). Critical of design technologies – particularly energy building instruments and quantitative simulations – they contend that subjective perceptions, through user experiences of buildings and subjective notions of space, are too often ignored. The result of Demers and Potvin's (2017) design and research speculations results in a melange that combines notions of biophilia and what they call material erosion. Consequently, their contribution to design knowledge regarding biophilic notions is primarily aesthetic:

> The design research innovates at three levels: its unique corpus of erosion typologies, its combined analogical and numerical simulation methodology and the representation of the dynamic nature of erosion into architectural forms.
>
> (Demers & Potvin, 2017, p. 340)

The idea that biophilia is employed by architects as an aesthetic also emerges in a comprehensive literature survey of biophilic design conducted by Wijesooriya and Brambilla (2020). In their survey of biophilic and sustainable design peer-reviewed literature, these authors understand that biophilic design can bridge the gap between quantitative and positivist notions of sustainable design (Wijesooriya & Brambilla, 2020). Wijesooriya and Brambilla (2020) contend that Green Building rating tools, and their reliance on technology solutions, lead to designs "that lack the human dimension' and these tools 'overlook the importance of connecting humans to nature" (p. 2). In a similar vein, like others, these authors claim that "human-nature connectedness" has real benefits for humans, and without this connection, humans may suffer "serious psychological difficulties" (Wijesooriya & Brambilla, 2020, p. 2). They support this assertion by pointing to the work of proponents of biophilic and restorative design in hospital design (Abdelaal & Soebarto, 2019).

Nevertheless, and despite the drift into unsubstantiated assertions, the survey produced by Wijesooriya and Brambilla (2020) is detailed and extensive. They chart the rise of Biophilic design in recent literature, identifying and examining 98 biophilic design articles between 2006 and 2019. A comprehensive top-down SWOT analysis was then used to examine the literature and is presented in a series of diagrams. As a result of this process 18 themes or "Strengths of adopting biophilic design within environmentally sustainable design" are named (Table 3, p. 6). The prevailing strengths of biophilia were identified as lower energy consumption and improved thermal comfort, yet most claimed benefits are naïve qualitative assertions. For example, aesthetic benefits such as the addition of "visual interest" and "natural beauty" (Wijesooriya & Brambilla, 2020, p. 6). Some of the predominant opportunities and attributes identified in the literature concern psychological biophilia, health and well-being, restorative pro-environmental behavior, and the enhancement of social interaction (Table 5, p. 9). These authors identify a range of research gaps in the biophilic design literature, such as research into biophilic education, policy studies, building performance and operations, and the integration of the biophilic into design processes (Wijesooriya & Brambilla, 2020, p. 12). They conclude in part that:

> It is clear, then, that the studies included in this review highlight the multi-disciplinary nature of BD and underline a significant research gap and lack of understanding of biophilia and its effect on humans, buildings and markets.
>
> (Wijesooriya & Brambilla, 2020, p. 12)

A lack of focus more generally is also noted in another recent survey of biophilic design in architecture. Zhong et al. (2021) conclude that:

> …generalised interpretations obscure the multiformity of biophilic design approaches and elements, which are obstacles to transforming this concept into practice. Furthermore, applying these elements call for knowledge from many subfields in architectural research, such as materiality, tectonics, mechanical systems, and mobility; however, these subfields are rarely linked in biophilic design studies. (p. 113)

In these surveys, it can be seen that biophilic design and its various tendencies do not directly address the climate emergency and carbon emissions.

Regenerative Design

Closely related to notions of biophilia is the push by some architects to adopt regenerative design principles. This model of architectural practice may well have, in future, much to offer sustainability transitions. This approach rejects technocratic models of eco-efficiency in favor of an approach that seeks to reconnect humans to the ecological processes of nature. The proponents of this approach argue that regenerative design allows for and requires multi-scalar, multi-layered, and interconnected gestures that continually build capacity within social-ecological systems. With its origins in the work of the Landscape architect John T Lyle who rejected linear systems in favor of self-generating and renewing cycles of circular flows (Lyle, 1996), regenerative design is also considered to address issues of culture, political agency, and human mindsets (Rojas, 2020, p. 99). For the most part, these principles, while well-meaning, are a syncretic amalgamation of various notions. Inevitably debates within architecture appear to center on definitions of sustainability and regeneration – should the latter replace the former. Moreover, the language of regenerative design often lapses into a series of essentialist aphorisms and generic prescriptions. Rhetoric centered on words such as cultivation, nurture, and nourishment pervading the regenerative discourse (Du Plessis & Brandon, 2015; Rojas, 2020, p. 212).

However, the language employed in these general prescriptions does not specifically address carbon emissions, certainly not in quantitative terms; it may be that a quantitative language is seen to be too much a mark of technocratic models to be spoken about. This language of good intentions often appears to be enough, and there are few studies on specific applications of regenerative design and its connection to policy interfaces. Any effort to transition to regenerative design principles, in whatever context, will inevitably result in issues of conflict between these principles and existing regimes of power. Without a shared language of political agency to address and work through policy conflicts, those architects that embrace the principles risk being isolated and cocooned in regenerative climate imaginaries.

As the above studies suggest, many architects are at best indifferent to and at worse complicit to the use of biophilia as a branding mechanism in the vortexes of property development. The contestation of biophilic principles in the media can be seen in the design of a new office building in Salford designed by Make as part of the New Bailey Project (Plot A3 New Bailey, Manchester, United Kingdom). The architects appear to have cloaked the building in a range of green adjectives: sustainability, holistic, low-energy, and renewable, in a mix that will "drive opportunity, growth, and prosperity for all" (Make, 2021). It was asserted that the building had met the UKGBC's net-zero carbon targets, and was close to meeting the WELL Building Standard principles, BREAM, as well as being a LETI Pioneer project (LETI, n.d; Make, 2021). LETI is the London Energy Transformation Initiative (LETI), a voluntary group of developers and AEC consultants established in 2017 to support the transition of London's buildings to Net Zero Carbon. Notions of Biophilia were at the heart of the Make design:

> ...applied an ivy veil with a mix of trailing and climbing plants, almost acting as a camouflage to the building as a whole.
>
> (Make, 2021)

This veil or wall was described in the following terms:

> This living wall will absorb and filter the pollution and carbon in the atmosphere, enhance biodiversity, attenuate rainwater, reduce noise, and provide biophilic benefits for the health and well-being of the users.
>
> (Make, 2021)

John Alker, director of policy and places at the UK Green Building Council (UKGBC), asserted that the building and the overall New Bailey development would:

> ...support the health and well-being of people and the nature and biodiversity of the surrounding area.
>
> (Make, 2021)

Prompted by the above claims, an article decrying biophilic or green office buildings – constructed from concrete, steel, and glass – Oliver Wainwright the Guardian's architecture critic, as well as Joe Giddings of ACAN, called out the New Bailey building as a sham: "the 'dirty secret' of similar green and so-called 'fossil-fuel free' office buildings" (Wainwright, 2021) (Fig. 5.1).

At this point, in the light of the above survey of biophilic tendencies in architecture, the way in which architects approach ecological science is worth briefly discussing. In *Constructed Ecologies* (2017), Margaret Grose, a landscape architect and ecologist, argues that landscape architects, as well as architects, require an intricate and more realistic understanding of the ecological sciences. Central to this argument is the idea that there is a "pressing need to distinguish fact from cultural opinions about the nature of Nature" (Grose, 2017, p. 3). Grose contends that designers need to recognize the difference between cultural knowledge, or "knowledge with a cultural overlay," and knowledge that is at the other extreme immutable because it is based on science (2017, p. 3). As she states, via this distinction:

> We can change policies, but we cannot ask plants to convert more sugars or produce more oxygen tomorrow or ask a tropical plant to live in the Sahara. Nature is not a human construct dependent on our own limitless imagination but is one of physical and finite riches. (2017, p. 3)

Grose's argument regarding "knowledge with a cultural overlay" speaks to the idea that for architects, there is the danger that scientific facts regarding the climate may be conflated with other ways of knowing. Biophilia and regenerative design have been read and transformed into projects in ways that are more general, cultural, and symbolic rather than reliant on ecological or medical science. Mazuch's (2017) earlier claimed that empirical "recent research" supported the benefits of biophilic design is not based on evidence-based research. But in fact

Fig. 5.1. Masterplanning for Anaklia Smart City (ASC), Georgia, Weston Williamson + Partners, Black Sea Coast, Anaklia Georgia. *Source*: Architectural Design. Burry (2020).

Frumkin (2011), the recent research cited by Mazuch, found that the evidence base supporting biophilic design was not well advanced. As the above survey suggests, for biophilia and regenerative design little may have changed. Both are generalized concepts that have been, and might too easily be, adopted by both architects and capital. More harshly: for architects, biophilic design has become a green petri dish, in which associated fragments of scientific facts, images, and cultural ideas are grown to produce a climate imaginary of the good Anthropocene.

Green Digital Cities

If the green cities require green materials and biophilic connectedness to come into being, then they also need an operating system. The digital city – and its variant the so-called smart city – has numerous proponents. These proponents see the digital city, the city of data, as the systemic network that will bring all the elements of life, both human and nonhuman – in fact of everything – together. This is the idea of the good Anthropocene writ large. In architecture, the proponents of the digital city since the *Non-Standard Architectures* exhibition have produced an impressive range of work, unbuilt projects, and texts. The idea of the digital city has influenced thinking within architecture schools becoming a dominating mode of pedagogy – and this is not to be churlish. As a social-technical imaginary, the digital city is seen as bringing together the infrastructure and instruments that will avert climate catastrophe. However, if we are to follow the lead of Nightingale et al. (2019), we can ask, what has this pursuit of the digital city foreclosed?

Mark Burry, a leading proponent of the digital city, writing in a special edition of the *Architectural Design* entitled *Urban Futures: Designing the Digitalised City* points to many of the familiar narratives that characterize the digital city perspective (Burry, 2020). These narratives all too often revolve around anthropocentric ideas of mastery and technical optimism. The first familiar trope of the argument is one of change: "cities are constantly changing" (p. 7). There is then a characteristic appeal to historic precedent drawn from the history of architecture: the Spanish engineer Ildefons Cerda, John Snow the English Physician, and the example of Ernst Fooks, an emigre architect working in Melbourne Australia in the 1950s. For Burry, the Anthropocene is the setting for the digitalized city, an unconstrained "opportunity" for innovation in architecture, cities, urbanism, and new data technologies. This is a perspective that has been stripped of any idea that we might embrace climate, environmentalism, or the material ontologies that seek to address nonhuman ecologies. What is espoused, seemingly oblivious of the debates across political ecology, is a political economy that seeks to accelerate – without saying it – capitalist growth and business as usual. The smart city "is both a challenge and an opportunity" in order to ensure "future prosperity" (Burry, 2020, p. 10). With data in hand, visionary and enlightened managerialism comes to the fore:

> The digitalised city needs to be designed: can the urban designer claim the conceit that this is naturally their task? If it is, how might they shape their role to facilitate the urban fabric's optimal evolution and link it with corresponding societal transitions?
>
> (Burry, 2020, p. 8)

Of course, this argument must counter an existing evil, and in this case, it is two-dimensional planning and optimized engineering thinking (Burry, 2020, p. 8). The next trope of this way of thinking is to advance the hyperbole of "visionaries" and "multi-disciplinary teams of urban futures analysts" who can bring different domains of knowledge together, and systems theory is often called upon to support this approach (Burry, 2020, p. 11). Despite this syncretic melange of preexisting tropes, there is an appeal to new ways of thinking – "Inferencing" as Burry names it – along with the invocation of popular science and even "NASA" so we can delve "deeply into the unknown unknowns of urban future experience" (2020, p. 11) (Fig. 5.2).

As practical examples of all this, Burry (2020) cites the example of two projects from Weston Williamson + Partners. One project is the Anaklia Smart City in

Fig. 5.2. Circular City, Weston Williamson + Partners, 2018. Image Courtesy of Weston Williamson + Partners. *Source*: Architectural Design. Burry (2020).

Georgia, on the Black Sea, and the other is a project for a Circular City, the latter appearing to resemble a contemporary version Étienne-Louis Boullée's ideal cities (Burry, 2020, pp. 11–12). These are representations of cities stripped of everything human and nonhuman except their digital skins. These cities are supported by a plethora of jargon with words like clusters, connected, infrastructure, smart, sustainable, radical, models, and concentric; words that evoke a cloak of architectural theory while also enabling the ultimate flexibility of image production.

In a similar vein, Kvan (2020) articulates the above position with a little more nuance. Using the 19th Century sanitation orientated zoning and epidemiological data of London as precedent, they argue that data are essential to the urban design of cities. In citing this precedent, Kvan calls for new transparency in design, arguing that architects, and the way architects are educated, should embrace urban and big data (Kvan, 2020). In other words, architects should develop expertise in data analytics (Kvan, 2020, p. 28). In this conception of the data-driven city, this is a city that is systematized through its data. All of these data can be used to save energy, create so-called "healthy neighborhoods," and promote walkability, physical exercise, and overall well-being (Kvan, 2020, p. 28). Citing Brenner, Kvan claims that data erode the boundaries between the urban and the rural, making the case that "cities are systems" and that "everything is potentially urban data" (2020, p. 29). For Kvan, these data are necessary to "facilitate the evolution of cities as resilient self-organizing systems" (p. 29). Kvan – like his mentor William J Mitchell – argues that the challenge for architects is to use data in integrative ways in order for architects not to lose their agency to the algorithmic mechanics of the data miners (Raisbeck & Neish, 2017).

The approach of the above systems protagonists also comes together in the idea of Complex Adaptive Systems (CASs) theory. This theory encapsulates a kind of global managerialism on a global scale and extends to settings beyond the cities of the Global North (Kvan & Karakiewicz, 2019a, 2019b). Benoit Mandlebrot's work, speeded up and animated, can be seen as the template for CASs theory. Complex systems discourse appears to mimic and has its sources in Batty's seductive agent-based simulations of cities such as London. These adaptive simulations both act as new instruments for understanding cities as well as suggesting what a city is. As claimed by its proponents, much of this discourse has precedents based on mathematics, systems theory, and the work of 1960s urban theorists such as Christopher Alexander and Kevin Lynch (Batty et al., 2019, p. 27). Thus digital computing allows for the analytical instruments of data to be conflated with a theoretical conception of the city (The University of Melbourne, 2015). CASs theory claims to link the physical with the biological, and to model cities as biological organisms from the bottom up rather than being designed from the top-down. Cities are conceived as "open-ended, coupled social, and infrastructural networks" (Batty et al., 2019, p. 27). This narrative has the advantage of conceivably bridging the gap between the local and larger scales of cities.

In CASs narratives, there is also an entire litany of claims and attributes that grant to cities a vitalism. The attributes include internal intelligence, complexity, the capacity of cities to evolve, open-ended growth, and the layering of new technologies over previous systems. These attributes allow cities to be apprehended as entities with increasing complexity and richness. In short, cities can be modeled as CASs because:

> These are (1) feedback and reciprocal causal relations, (2) nonlinearity and thresholds, (3) historical path dependence and time lags, (4) resilience, and (5) heterogeneity.
>
> (Batty et al., 2019, p. 33)

Hence, CASs cities, as networked and dynamic systems, can easily be likened to ecosystems. Human agents are "embedded in networks" and the networks exhibit strong heterogeneity and diversity because "of the interconnection" of these different agents. Cities, like Mandlebrot's patterns, can be likened to "rain forests or coral reefs":

> This is a special feature of complex adaptive systems – latent also in very productive ecosystems, as in a rain forest or coral reefs – where specialisation and interdependence are immense and mesmerising.
>
> (Batty et al., 2019, p. 30)

This dream of the city as a system is an old dream, and it is one that appears again and again in architectural discourse. It is a dream easily linked to ideas of a self-organizing and living green city. Yet the polemics and impetus of these ideas have produced an apparatus within architecture, discourse, and education. In turn, this has produced various multiplier effects in architectural theory, education, and practice. Yet paradoxically, through its own framing, orientated to digital data, CASs theory might also be seen as actively playing a part in the co-production of climate change and perpetuating ecological violence. To return to Nightingale et al. (2019):

> How the climate problem is framed opens our imagination to seeing climate change as interlinked problems that permeate different scales and entangle socionatures in new ways. (p. 6)

There are no entangled socionatures in the above framing of CASs and big data in the city; the interlinking of ideas is too neat a solution; it is systems begetting systems, cybernetics creating cybernetics. It is a verbiage that has edited out any thought of political ecology. The big data mantra and the associated smart city appears to elide advances in digital ethnographies, the rights to the city, surveillance of citizens, gender and safety in cities, LGBTIQ spatiality, nor anything much about the Global South, and certainly the realities of climate change itself.

Conclusion: The Good Anthropocene

For many architects, the concept of the good Anthropocene, exemplifying a seemingly good postclimate emergency earth, has become the default position. The alternative to this dangerous idea is to recognize the climate emergency and pursue practices grounded in radical environmental politics and theories more fully. While numerous scholars have interpreted the Anthropocene differently – with all its different guises and dangers – architects have rarely been located within these debates (Sklair, 2017). Fremaux and Barry (2019) call the Anthropocene "a new signifier" that has generated "mastery and technological optimism in the face of environmental uncertainty" (Fremaux & Barry, 2019, p. 171). This observation, while it may go without saying, certainly accords with the practices of many architects described in this chapter. Fremaux and Barry (2019) identify three primary debates related to the Anthropocene and ideas of entanglement in various socio-natures; as they term it, the "intertwinement of nature and society around us – and inside us." (p. 171). These three trajectories are succinctly identified as follows:

> [Firstly] the promotion of techno-hybrid ontologies by neo-materialists (Barad, 2007; Bennett, 2010; Braidotti, 2013), [Secondly] techno-optimistic talk of a "good Anthropocene" from "ecomodernists" (Arias-Maldonado, 2015; Ellis, 2011; Nordhaus &

> Shellenberger, 2013), and [Thirdly] eco-managerial aspirations of earth system stewardship and governance.
>
> (Fremaux & Barry, 2019, p. 171)

These authors then map these three trajectories across two scales. To then develop four categories describing different responses to the Anthropocene concept. On one scale, they describe "Mostmodernity" versus "Postmodernity"; an arrogant domination of nature versus a prudent approach to nature. On the other scale, they describe "Humanism" versus "Posthumanism." This scale measures the degree to which humanist ideas treat the epistemological dualism of society and nature – greater entanglement being positive. These scales then generate four different categories as set out below.

First, the "Humanist mostmodern narrative" is elaborated as an "unreflexive humanism techno optimism." This is a business as usual narrative through which science acts on inert "nonhuman matter" and where:

> …scientists, considered as the new shepherds of humanity for having developed a global environmental consciousness and for planning to 'save the planet' thanks to more science and green geo-technologies. (p. 173)

Examples of this approach in architecture might be the work of BIG architects, Foster and Partners or Patrik Schumaker of Zaha Hadid architects – lesser known architects such as Make and William Weston would also fall into these categories.

Second, the "Posthumanist mostmodern [post-nature]" narrative is articulated as a "techno reproduction of nature." This is described as a position that:

> This post-nature narrative seeks to overcome the fake distinctions inherent in the modern project, such as the one between nature and culture but eventually, it overcomes the dualist ontology only to offer a techno-nature monism that reduces nature to a scientific and technological by-product of society. (p. 174)

The proponents of CASs would align with thus narrative. In other words, this is a position that:

> …while ostensibly upholding a decentered perspective (humans being themselves hybrid realities), it remains caught in the very modern Promethean conceit that nature can be controlled, made and it urges the production/reproduction of nature through technological artefacts. (p. 174)

Ecologic, Menges, and the Urbach Tower would fall into this category – as discussed in Chapter 2 – as well as the floating projects discussed in the previous chapter.

Third, on the "Postmodernity," more negative side of the domination of nature, there is the "Humanist postmodern narrative" explained as a "reflexive humanism – modesty and prudence – human centeredness." This is characterized by a "weak anthropocentrism" and this mode:

> ...works on the political and ethical reappraisal of modernity, the realisation of its emancipatory ideals while keeping epistemologically and ethically (but not ontologically) the human subject in the centre. Weak anthropocentrism, for instance, suggests that one can be human-centred without being focused only on human self-interest. (pp. 174–175)

Arguably this is the narrative adopted by biophilic and regenerative adherents. But it also characterizes various architectural theorists such as Spencer and Deamer, as discussed in Chapter 7, who reappraise modern architecture by turning to critiques of capital and labor that do not eschew human agency.

Finally, the "Posthumanist postmodern" narrative is enunciated through a "Connection with nature [and] Relational ontology." This narrative is established through positions of radical ecology and ecofeminism and is suggested in the work of Plumwood and Guattari and more closely aligned with Frichot, as discussed in Chapters 1 and 2. Hamilton et al., (2015, p. 280) and Fremaux and Barry (2019) contend this position considers the Anthropocene:

> ...(and the disappearance of nature) as both unjust and unsustainable, linked to the domination of the North and oriented towards the sustainability of a capitalist "world-system" within the earth system. (p. 175)

A position, more aligned with this book, that calls into question:

> ...the possibility for us, for other species and further generations, to survive or more precisely to survive in decent and humane conditions.
>
> (Fremaux & Barry, 2019, p. 176)

Indeed, many of the green projects of this chapter, and the biophysical projects of the previous chapter, can be characterized by a lack of hubris and an overdetermined managerialism. To adapt Fremaux and Barry's (2019) words: a new project of rethinking architecture's entwinement with nature in the Anthropocene is needed. And this project would effectively eschew human hubris and the overriding desire to control "the Earth itself and its various biophysical entities and processes" (p. 7). This process of rethinking architecture's relationship with the Earth would view humanity's relationship with the Earth as inherently political (p. 2). As they write:

> Rather, and here exhibiting long- established green political ideas, *what is in need of management is humanity's relationship with the Earth.* This clear focus on human self-governance brings green politics (and ethics) "back in," thus countering the depoliticised and depoliticising arguments of "the good Anthropocene." (p. 2)

Only a few architects and theorists appear to be pursuing the green politics called for by Fremaux and Barry (2019). Even those architectural theorists who radically question binary notions of society and nature are too often mired in ontologies that give primacy to matter, without proposing a broader reworlding of the Earth.

Given all of the above, it is worth returning to Rossi as a kind of talisman to see what their distant work might then suggest about the present. First, Rossi's approach developed out of a milieu in Italian architecture riven by a heterogeneity of different factions and debates. This heterogeneity of theory, positions, and debate, is not immediately evident at present in the discourse around sustainable architecture. There is no sense of fear driven by climate urgencies in these debates – and perhaps there should be. In this discourse, the undifferentiated goal of the good Anthropocene dominates. The green city all too often is rendered as a holistic yet monolithic entity. This is an imagined city that has no need for climate realism and scarcely little concern for climate science. This is an idea of the city that fails to confront the realism of the Earth's tipping points or ecocide. In looking outside of architecture, Rossi's encounter with French geography and urban science was not riven by a similar fabulism (Rossi, 1982). In short, it was not an attempt to link architecture to sciences, far removed from the instruments of architecture itself. Nor was Rossis' encounter an effort to draw on the miasma of popular science. Rossi's approach – his neorealism if we are to call it that – was not to see the city as a system divorced from the contingencies and accidents of history. In addition, for all its poetics, Rossi's work expanded and linked to the collective memory of cities and their spaces. These urban spaces, like the Floating Theater at Venice, suggested the possibility of events, rather than an aphasia that never confronts either a liberatory poetics or politics.

The green city, implied in many of the projects and theories, in this chapter is an entity that is cloaked in the image-politics of optimism. The technologies of design and governance of this city appear morally precise. Yet, they are easily distributed, entirely adaptable to any place, easily scaled up or down, and comfortably claimed by regimes of power. This green city of good virtues, of a virtuous Anthropocene, forecloses the possibility of a radical climate poetics that is simultaneously aesthetic and political. In doing so, this city can only be a ship of fools, a funereal barge, devoid of a politicized poetics, forever changing and growing, with architects at the tiller, leading to the Earth's downfall.

References

Abdelaal, M. S., & Soebarto, V. (2019). Biophilia and Salutogenesis as restorative design approaches in healthcare architecture. *Architectural Science Review*, *62*(3), 195–205.

Ade, R., & Rehm, M. (2020). The unwritten history of green building rating tools: A personal view from some of the 'founding fathers'. *Building Research & Information*, *48*(1), 1–17. doi:10.1080/09613218.2019.1627179

Alla, H., & Khalil, E. E. (2020). Enhancing livability through resource efficiency: An urban metabolism study in Cairo. In I. Ruby & A. Ruby (Eds.), *The materials book* (pp. 191–197). Berlin: Ruby Press.

Angélil, M., & Siress, C. (2020). The matter of construction: Systemic overhaul or tweaking the status quo? In I. Ruby & A. Ruby (Eds.), *The materials book* (pp. 9–24). Berlin: Ruby Press.

Arias-Maldonado, M. (2015). *Environment and society: Socionatural relations in the Anthropocene*. Berlin: Springer.

Barad, K. (2007). *Meeting the universe halfway*. Durham, NC: Duke University Press.

Batty, M., Bettencourt, L. M., & Kirley, M. (2019). Understanding coupled urban-natural dynamics as the key to sustainability: The example of the Galapagos. In T. Kvan & J. Karakiewicz (Eds.), *Urban Galapagos: Transition to sustainability in complex adaptive systems* (pp. 23–41). Berlin: Springer.

Bennett, J. (2010). *Vibrant matter: A political ecology of things*. Durham, NC: Duke University Press.

Braidotti, R. (2013). *The posthuman*. Cambridge: Polity.

Braungart, M., & McDonough, W. (2009). *Cradle to cradle*. New York, NY: Random House.

Burry, M. (2020). Better to make a good future than predict a bad one. *AD Special Issue: Urban Futures: Designing the Digitalised City*, *90*(3), 6–13. Retrieved from https://onlinelibrary.wiley.com/doi/10.1002/ad.2561

Business & Human Rights Resource Centre (BHRRC). (n.d). Esquel Group denies allegations of using forced labour in Xinjiang and appeals against US sanction. Retrieved from https://www.business.humanrights.org/en/latest-news/esquel-group-deniesallegations-of-using-forced-labour-in-xinjiangand-appeals-against-us-sanction/retrieved211021

Clark, H., & Tilly, M. (2021, December 1). The curious case of Scarborough emissions. *Energy News Bulletin*. Retrieved from https://www.energynewsbulletin.net/development/news/1422564/the-curious-case-of-scarborough-emissions

Coxe, W., Hartung, N. F., Hochberg, H. H., Lewis, B. J., Maister, D. H., Mattox, R. F., & Piven, P. A. (1987). *Success strategies for design professionals*. New York, NY: McGraw-Hill.

Crawford, R. H., Bontinck, P. A., Stephan, A., Wiedmann, T., & Yu, M. (2018). Hybrid life cycle inventory methods–a review. *Journal of Cleaner Production*, *172*, 1273–1288. doi:10.1016/j.jclepro.2017.10.176

Crawford, R. H., Stephan, A., & Prideaux, F. (2020, November). EPiC: Introducing a database of hybrid environmental flow coefficients for construction materials. *IOP Conference Series: Earth and Environmental Science*, *588*, 42052.

Crawford, R. H., Stephan, A., & Prideaux, F. (2021, November). The EPiC database: Hybrid embodied environmental flow coefficients for construction materials. *Resources, Conservation and Recycling*, 1061058. doi:10.1016/j.resconrec.2021.106058

Crawford, R. H., Stephan, A., & Schmidt, M. (2018). Embodied carbon in buildings: An Australian perspective. In F. Pomponi, C. De Wolf, & A. Moncaster (Eds.), *Embodied carbon in buildings* (pp. 393–416). Berlin: Springer.

Crook, L. (2021, April 22). Ten ways in which architecture is addressing climate change. *Dezeen*. Retrieved from https://www.dezeen.com/2021/04/22/architecture-climate-change-earth-day/

DeLuca's. (1999/2005). *Image politics: The new rhetoric of environmental activism*. New York, NY: Guildford Press.

Demers, C. M., & Potvin, A. (2017). Erosion in architecture: A tactile design process fostering biophilia. *Architectural Science Review*, *60*(4), 325–342.

Design Anthology. (n.d). A textile factory that weaves together sustainability and industry. Retrieved from https://design-anthology.com/story/integralretrieved 211021

Design Awards. (2021). The Kendeda building for innovative sustainable design [webpage]. Retrieved from https://www.dezeen.com/awards/2021/shortlists/the-kendeda-building-for-innovative-sustainable-design/retrieved211021

Dezeen Awards. (2021a). Longlists-sustainability. Dezeeen. Retrieved from https://www.dezeen.com/awards/2021/longlists/sustainability/

Dezeen Awards. (2021b). Olympic house-IOC headquarters by 3XN. Dezeen. Retrieved from https://www.dezeen.com/awards/2021/longlists/olympic-house-ioc-headquarters/

Dezeen Awards. (2021c). Catalyst building by MGA Michael green architecture. Retrieved from https://www.dezeen.com/awards/2021/longlists/catalyst-building/

Dezeen Awards. (2021d). Integral by Ronald Lu & partners [webpage]. Retrieved from https://www.dezeen.com/awards/2021/longlists/integral-2/retrieved211021

Dezeen. (n.d). About. Retrieved from https://www.dezeen.com/about/retrieved211021

Du Plessis, C., & Brandon, P. (2015). An ecological worldview as basis for a regenerative sustainability paradigm for the built environment. *Journal of Cleaner Production*, *109*, 53–61.

Ellis, E. C. (2011). Anthropogenic transformation of the terrestrial biosphere. *Philosophical Transactions of the Royal Society A*, *369*(1938), 1010–1035.

Esquel Group. (2020, September 2). Correcting the record about Esquel's presence and operations in Xinjiang introduction to Esquel group and our values [webpage]. Retrieved from https://www.esquel.com/news/correcting-record-about-esquel%E2%80%99s-presence-and-operations-xinjiang

Esquel Group. (n.d). Our history. Retrieved from https://www.esquel.com/our-history

Fairs, M. (2021, 24 August). Concrete construction 'offsets around one half' of emissions caused by cement industry says IPCC. Dezeen. Retrieved from https://www.dezeen.com/2021/08/24/concrete-construction-offsets-emissions-cement-industry-ipcc/

Fremaux, A., & Barry, J. (2019). The 'good Anthropocene' and green political theory: Rethinking environmentalism, resisting eco-modernism. In F. Biermann &

E. Lövbrand (Eds.), *Anthropocene encounters: New directions in green political thinking* (pp. 171–190). Cambridge: Cambridge University Press.

Frumkin, H. (2011). Nature contact and human health: Building the evidence base Howard Frumkin. In S. R. Kellert, J. Heerwagen, & M. Mador (Eds.), *Biophilic design: The theory, science and practice of bringing buildings to life* (pp. 107–118). New York, NY: John Wiley & Sons.

Fu, X., & You, W. (2018). Basic green building design: Reconnecting sustainability to the vernacular. *Architectural Design*, *88*(6), 80–87.

Garber, R. (2017). Ecological workflows: Zhangdu Lake Farm, Hubei Province, China. *Architectural Design*, *87*(3), 114–119.

Garcia-Lamarca, M., Anguelovski, I., Cole, H., Connolly, J. J., Argüelles, L., Baro, F., … Shokry, G. (2021). Urban green boosterism and city affordability: For whom is the 'branded' green city? *Urban Studies*, *58*(1), 90–112. doi:10.1177/0042098019885330

Gray, S. G., Sütterlin, B., Siegrist, M., & Árvai, J. (2020). The benefit of virtue signaling: Corporate sleight-of-hand positively influences consumers' judgments about "social license to operate". *Journal of Environmental Management*, *260*(April), 110047. doi:10.1016/j.jenvman.2019.110047

Grose, M. (2017). *Constructed ecologies: Critical reflections on ecology with design.* New York, NY: Taylor & Francis.

Guallart, V. (2020). From digital cities to Biocities: Harnessing the power of the digital revolution to reinvent the urban ecology model. *Architectural Design*, *90*(3), 72–75. doi:10.1002/ad.2571

Hamilton, C., Bonneuil, C., & Gemenne, F. (Eds.). (2015). *The Anthropocene and global environmental crisis.* London: Routledge.

Hariman, R., & Lucaites, J. L. (2019). From image politics to image politics 2.0. *Communication and Critical*, *16*(4), 340–349.

International Living Future Institute. (n.d). FAQ. Retrieved from https://living-future.org/contact-us/faq/

ISO, I. (2006). ISO-14040 environmental management–life cycle assessment–principles and framework. International Organization for Standardization. Retrieved from https://www.iso.org/standard/37456.html

Joshi, S. (1999). Product environmental life-cycle assessment using input-output techniques. *Journal of Industrial Ecology*, *3*(2), 95–120.

JP/Politiken Media Group. (n.d). Key figures. Retrieved from https://jppol.dk/en/om-os/noegletal-og-aarsrapporter/

Kellert, S. R., Heerwagen, J., & Mador, M. (2011). *Biophilic design: The theory, science and practice of bringing buildings to life.* New York, NY: John Wiley & Sons.

Kellert, S. R., & Wilson, E. O. (1993/2013). *The biophilia hypothesis.* Shearwater. Washington, D.C.: Island Press.

Klöpffer, W. (2014). Introducing life cycle assessment and its presentation in 'LCA compendium'. In W. Klöpffer (Ed.), *Background and future prospects in life cycle assessment* (pp. 1–37). Dordrecht: Springer.

Kvan, T. (2020). Data-informed design: A call for theory. *Architectural Design*, *90*(3), 26–31.

Kvan, T., & Karakiewicz, J. (2019a). Complexity and consequence in coupled natural urban systems. In T. Kvan & J. Karakiewicz (Eds.), *Urban Galapagos: Transition to sustainability in complex adaptive systems* (pp. 1–6). Berlin: Springer.

Kvan, T., & Karakiewicz, J. (Eds.). (2019b). *Urban Galapagos: Transition to sustainability in complex adaptive systems*. Berlin: Springer.

LETI. (n.d). About. Retrieved from https://www.leti.london/about

Lorch, R. (2017). Climate policies for buildings. *Building Research & Information, 45*(5), 475–477.

Lu, A., & Dou, P. (Eds.). (2018). China homegrown: Chinese experimental architecture reborn. *Architectural Design, 88*(6), 1–146.

Lyle, J. T. (1996). *Regenerative design for sustainable development*. New York, NY: John Wiley & Sons.

Make. (2021, February 8). Striking commercial scheme with living wall at New Bailey is given green light by Salford planners. Retrieved from https://www.makearchitects.com/media/news-events/striking-commercial-scheme-with-living-wall-at-new-bailey-is-given-green-light-by-salford-planners/

Marcus/Dezeen. (2021, August 27). Concrete construction 'offsets around one half' of carbonate emissions from cement industry says IPCC. Dezeen. Retrieved from https://www.dezeen.com/2021/08/24/concrete-construction-offsets-emissions-cement-industry-ipcc/

Mazuch, R. (2017). Salutogenic and biophilic design as therapeutic approaches to sustainable architecture. *Architectural Design, 87*(2), 42–47.

Mikaela. (2021, August 26). Concrete construction 'offsets around one half' of carbonate emissions from cement industry says IPCC. Dezeen. Retrieved from https://www.dezeen.com/2021/08/24/concrete-construction-offsets-emissions-cement-industry-ipcc/. Accessed on June 9, 2021.

Minunno, R., O'Grady, T., Morrison, G. M., & Gruner, R. L. (2021). Investigating the embodied energy and carbon of buildings: A systematic literature review and meta-analysis of life cycle assessments. *Renewable and Sustainable Energy Reviews, 143*(June), 110935. doi:10.1016/j.rser.2021.110935

Mohebbi, G., Bahadori-Jahromi, A., Ferri, M., & Mylona, A. (2021). The role of embodied carbon databases in the accuracy of life cycle assessment (LCA) calculations for the embodied carbon of buildings. *Sustainability, 13*(14), 7988. doi:10.3390/su13147988

Morris, F., Allen, S., & Hawkins, W. (2021). On the embodied carbon of structural timber versus steel, and the influence of LCA methodology. *Building and Environment, 206*(108285). Retrieved from https://www.sciencedirect.com/science/article/abs/pii/S0360132321006843?via%3Dihub

Nichols, S. (2018). Pollux's spears. *Grey Room*, 71, 141–155. doi:10.1162/grey_a_00246

Nightingale, A. J., Eriksen, S., Taylor, M., Forsyth, T., Pelling, M., Newsham, A., … Whitfield, S. (2019). Beyond technical fixes: Climate solutions and the great derangement. *Climate & Development, 12*(4), 343–352.

Niskanen, J., & Rohracher, H. (2020). Passive houses as affiliative objects: Investment calculations, energy modelling, and collaboration strategies of Swedish housing companies. *Energy Research & Social Science, 70*, 101643. doi:10.1016/j.erss.2020.101643

Nordhaus, T., & Shellenberger, M. (2007). *Break through: From the death of environmentalism to the politics of possibility*. Boston, MA: Houghton Mifflin.

Pfannes, E. (2020). Paradigm shift: The city of 1,000 talks, Chennai. In I. Ruby & A. Ruby (Eds.), *The materials book* (pp. 169–180). Berlin: Ruby Press.

Pryshlakivsky, J., & Searcy, C. (2013). Fifteen years of ISO 14040: A review. *Journal of Cleaner Production*, *57*, 115–123.

Raisbeck, P. (2019). *Architecture as a global system: Scavengers, Tribes, Warlords and Megafirms*. Bingley: Emerald Publishing Limited.

Raisbeck, P., & Neish, P. (2017). Digital fragments and historiographies. In Quotation, quotation: Proceedings on the 34th annual conference of the society of architectural historians, Australia and New Zealand, SAHANZ.

Ramzy, N. S. (2015). Biophilic qualities of historical architecture: In quest of the timeless terminologies of 'life' in architectural expression. *Sustainable Cities and Society*, *15*(July), 42–56. doi:10.1016/j.scs.2014.11.006

Ravenscroft, T. (2019, June 19). 3XN builds new headquarters for the Olympics on shore of Lake Geneva. Dezeen. Retrieved from https://www.dezeen.com/2019/06/19/olympic-house-3xn-lake-geneva-lausanne-switzerland/retrieved211021

Ravenscroft, T. (2021, March 9). Dezeen acquired by JP/Politiken media group. Dezeen. Retrieved from https://www.dezeen.com/2021/03/09/dezeen-acquired-jp-politiken-media-group/

Röck, M., Saade, M. R. M., Balouktsi, M., Rasmussen, F. N., Birgisdottir, H., Frischknecht, R., … Passer, A. (2020). Embodied GHG emissions of buildings–the hidden challenge for effective climate change mitigation. *Applied Energy*, *258*. doi: 10.1016/j.apenergy.2019.114107

Rojas Gracia, A. (2020). Transformative processes for architectural design: A heuristic study of regenerative practices. Doctoral dissertation.

Rossi, A. (1982). *The architecture of the city* (P. Eisenman, Trans.). Cambridge, MA: MIT Press.

Ruby, I., & Ruby, A. (2020). *The materials book*. Berlin: Ruby Press.

Samuel, F. (2018). *Why architects matter: Evidencing and communicating the value of architects*. London: Routledge.

Schlögel, R., Kofler, C., Gariano, S. L., Van Campenhout, J., & Plummer, S. (2020). Changes in climate patterns and their association to natural hazard distribution in South Tyrol (Eastern Italian Alps). *Scientific Reports*, *10*, 5022. doi:10.1038/s41598-020-61615-w

Sklair, L. (2017). Sleepwalking through the Anthropocene. *British Journal of Sociology*, *68*(4), 775–784. doi:10.1111/1468-4446.12304

Suh, S., & Huppes, G. (2002). Missing inventory estimation tool using extended input-output analysis. *International Journal of Life Cycle Assessment*, *7*(3), 134–140.

Tafuri, M. (1989). *History of Italian architecture, 1944–1985*. Cambridge, MA: MIT Press.

The University of Melbourne: AURIN. (2015, November 10). Future cities: The good, the bad and the ugly: Michael Batty [video]. YouTube. Retrieved from https://www.youtube.com/watch?v=z1hsbyqyf8k

Tilly, M. (2021, October 7). XR protestors glue themselves to Woodside HQ. *Energy News Bulletin*. Retrieved from https://www.energynewsbulletin.net/environment/news/1419116/xr-protestors-glue-themselves-to-woodside-hqHQ

Treloar, G. J. (1997). Extracting embodied energy paths from input–output tables: Towards an input–output-based hybrid energy analysis method. *Economic Systems Research*, *9*(4), 375–391.

Wainwright, O. (2021, April 3). Green building—The dirty secret of so-called 'fossil fuel free' buildings. *The Guardian*. Retrieved from https://www.theguardian.com/environment/2021/apr/03/dirty-secret-fossil-fuel-free-building-embodied-carbon-building-glass-steel-blocks

Wallace, E., Buil, I., & De Chernatony, L. (2020). 'Consuming good' on social media: What can conspicuous virtue signaling on Facebook tell us about prosocial and unethical intentions? *Journal of Business Ethics*, *162*(3), 577–592. doi:10.1007/s10551-018-3999-7

Wijesooriya, N., & Brambilla, A. (2020). Bridging biophilic design and environmentally sustainable design: A critical review. *Journal of Cleaner Production*, *283*, 124591. doi:10.1016/j.jclepro.2020.124591

Wilson, E. O. (1986). *Biophilia*. Cambridge, MA: Harvard University Press.

Xiuzhong Xu, V., Cave, D., Leibold, J., Munro, K., & Ruser, N. (2020, March 1). Uyghurs for sale–'re-education', forced labour and surveillance beyond. Australian Strategic Policy Institute. Retrieved from https://www.aspi.org.au/report/uyghurs-sale

Zad de la Colline. (2020, October 25). ZAD de la colline: Orchidées contre béton armé! [video]. YouTube. Retrieved from https://www.youtube.com/watch?v=JKwM50uHiB4&t=498s

Zad de la Colline. (n.d). Zad de la Colline. Retrieved from https://zaddelacolline.info/en.html

Zhong, W., Schröder, T., & Bekkering, J. (2021). Biophilic design in architecture and its contributions to health, well-being, and sustainability: A critical review. *Frontiers of Architectural Research*, *11*(1), 114–141. doi:10.1016/j.foar.2021.07.006

Chapter 6

Environmental Politics and Design Activism

Introduction

To move now in this chapter to exploring the degree to which architects act as climate activists dedicated to averting dangerous climate change. Central to exploring this issue is to ask, to what extent are architects involved in civil society? In the aftermath of COP26, advocacy is ever more urgent to achieve no more than – and indeed less than – 1.5 °C of global warming. COP26 did not close the 1.5 °C emissions gap, but it put in place a process that might enable this gap to be closed. While many countries have adopted net-zero policies for 2050 and beyond, this may not be enough, and global emissions must be reduced quickly and urgently by 2030. For architects, the civil society groups and climate activists that surround the COPs, the UNFCCC and The UN-Habitat system point to the global institutions that architects might, and must by necessity, engage with to forestall the Earth's temperature rising above 1.5 °C. At a national level architects, through their existing membership bodies, but more so with the recent emergence of climate groups such as Insulate Britain and Extinction Rebellion, have an opportunity to engage with ecopolitics. At the end of COP26, the COP26 President Rt Hon Alok Sharma MP summarized the situation:

> We can now say with credibility that we have kept 1.5 degrees alive. But, its pulse is weak and it will only survive if we keep our promises and translate commitments into rapid action. We must now move forward together and deliver on the expectations set out in the Glasgow Climate Pact. It is up to all of us to sustain our lodestar of keeping 1.5 degrees within reach and to continue our efforts to get finance flowing and boost adaptation. After the collective dedication which has delivered the Glasgow Climate Pact, our work here cannot be wasted.
>
> (United Nations Climate Change, 2021, p. 2)

COP26, as many have noted, was to be the last chance for nations to avert dangerous climate change by keeping global warming to 1.5 C. If more advocacy is needed to do this, and COP26 is only a waypoint after the 2015 Paris Agreement, one must now ask what roles do architects have to play as global activists in order

Architects, Sustainability and the Climate Emergency, 155–188

Published under exclusive licence by Emerald Publishing Limited
doi:10.1108/978-1-80382-291-420221006

to close this gap? For some commentators, COP26 was seen as the first COP where the physical science was recognized and accounted for; given the discussions in the previous chapters, this raises a second question for architects: To what extent do architects embed the physical science of climate into their own calculations, theories, research, spatial practices, projects – small and large – and advocacy.

In a study of COP21 Guillermot (2017) notes that the connections between "science and politics are more complex, and the boundaries between them more mobile and negotiated" (p. 41). While this point may seem obvious, Guillemot (2017) extends it by arguing that at the time of the creation of the UNFCCC in 1982 and the IPCC in 1988, the relationship between science and politics was a direct "linear model" (p. 41). This linear model equated to scientists providing scientific knowledge to politicians who then acted independently to make policy decisions. The Copenhagen COP challenged this linear viewpoint, and the widely accepted scientific consensus on Climate Change was "not sufficient to trigger political action" (p. 41). By the time of COP21 in Paris, the 1.5 °C target adopted was a more stringent target than what had previously been adopted via the previous 2 °C target (Guillemot, 2017, p. 2).

It is easy to be confused about strategies and tactics for insurrection, protest, and alternative movements. All the more so in the aftermath of the Occupy movement, the tragedy of the 2011 Egyptian revolution, the lessons of Extinction Rebellion, and the Blacks Lives Matter movement. George Lakey, a peacemaker and radical activist, proposes five stages for protest movements (Lakey, 1973). As set out in Table 6.1, these stages have been adapted for architectural organizations – now making seven stages – and this is set out in Table 6.1. The stages employed here are: Membership Aggregator (MA), Cultural Preparation (CP), Organisation Building (OB), Confrontation (Con), Mass Non-cooperation (MN-c), Parallel Institutions (PI), and Not Evident (NE).

In Lakey's model, Cultural Preparation (CP) includes analysis, visioning, and strategy formulation. Lakey makes the point that problems of analysis may emerge quickly. This raises the issues for architects whether or not the the profession can quickly respond to fast-moving events. In the second stage, the stage of Organisation Building (OB), there is an emphasis on building a nonhierarchical organization with distributed models of leadership. This is in order to avoid the problems associated with developing an organizational bubble or cult mentality. In the Confrontation stage (Con), Lakey argues that this stage will not be effective without the implementation of the prior stages. In the next stage, Lakey proposes the need for peaceful mass Non-Cooperation (Mn-c), which results in the final stage of developing alternative and Parallel Institutions (PIs). Perhaps Lakey's framework may seem too simplistic. Nonetheless, this framework is a useful way to benchmark the success and evolution of particular protest movements.

It is, therefore, reasonable to use Lakey's framework to understand how architects produce political ecologies in their own discourse (American Friends Service Committee, 2012; Lakey, 1973). Lakey's framework is a benchmark for considering, coming as it does from the activism of the 1960s, how various architectural groups have formed themselves to constitute a part of civil society. The events of May 1968 have appeared to have been a beacon of deep nostalgia

Table 6.1. Architectural Advocacy Groups.

Architectural Organization	Transparent Governance	International Reach	Protest Life Cycle	Education Programs	Presence in UNFCCC or UN-Habitat	Funding and Sponsors	Web References
Africa Union of Architects /Union des Architectes d'Afrique (AUA, UAA)	Nongovernmental organization connected to key professional organizations in 41 countries. (e.g., South Africa, Nigeria, and Tunisia).	Africa (Founded in Lagos Nigeria, 1981). Platform	MA: African Architects and design and construction professionals.	NE	UN Habitat.	NE	https://uia.org/s/or/en/1100059167; https://www.facebook.com/aua.org/.
American Institute of Architects AIA (USA)	Membership organization governed by a Board of Directors. Works with private/corporate partners. Over 200 chapters in the United States and internationally.	Nationally orientated but membership open to professionals outside the United States. (Founded in New York, 1857).	MA, CP, and OB: Advocacy for profession includes campaigns and lobby's on policy to legislators at the federal, state, and local level.	Yes, professionally orientated.	No.	Members are predominately architects and built environment professionals. Private Industry strategic and collaboration partners.	https://www.aiainternational.org/; See climate action program: https://www.aia.org/resources/77541-where-we-stand-climate-change.
Architects Climate Action Network (ACAN)	Organization defines itself as a political organization and "a network of	National (UK) with loose global connections.	MA, CP, OB, and C: Climate Advocacy and protest activities include political	Yes. Professionally orientated.	Yes. Presence and programme of events at COP26.	Members are predominately architects and built	https://www.architectscan.org/.

Table 6.1. *(Continued)*

Architectural Organization	Transparent Governance	International Reach	Protest Life Cycle	Education Programs	Presence in UNFCCC or UN-Habitat	Funding and Sponsors	Web References
	individuals within architecture and related built environment professions" (ACAN, n.d.a).		campaigning and lobbying, and educational reform.			environment professionals.	
Architects Council of Europe /Conseil Des Architectes D'Europe (ACE)	Nonprofit. Leadership: General assembly composed of a delegation for each member organization. President, five-member executive board, and general secretariat 45 Member Organizations across the EU Member States representing the interests of over 600,000 architects.	Europe (Founded in Treviso, Italy, 1990).	MA and CP: General advocacy for the profession.	Yes, professionally orientated.	No.	45 Member organisations, which are the regulatory and professional representative bodies in all EU Member States, UK, Switzerland and Norway.	https://www.ace-cae.eu/. Note: Sustainable Architecture and Urban Issues (ESA-UI) Work Group. See: *The Role of the Architectural Profession in Delivering Responsible Design, 2014* (ACE, 2014). Supporting Sustainable Development of the Built Environment. For more see: https://www.ace-cae.eu/about-us/mission-and-objectives/.
Architecture 2030	Nonprofit, 501(c)(3) research	US focus with global reach	MA, CP, and OB: Climate		No. Architecture 2030 issued the 2030	Foundation grants or	https://architecture2030.org/.

	organization (USA). Nonpartisan organization. Leadership, founder, and CEO; president; senior program director; as well as senior fellows and board members.	(Founded 2002).	orientated advocacy aimed at profession and related built industries.	Yes, professionally orientated.	Challenge adopted by the AIA in 2006	contributions from architecture firms, corporate sponsors as well as individual donations.	
Architects Declare (Australia)	Signatory organization of a network of architectural practices and steering group members.	National with loose global connections (Launched, 2019).	MA, CP, and OB: Industry recognized declaration with signatories.	Provides resources.	Minimal.	Members are predominately architects and built environment professionals.	https://au.architectsdeclare.com/.
Architects Declare (UK)	Signatory organization of a network of architectural practices and steering group members.	National with loose global connections (Launched, 2019).	MA, CP, and OB: Industry recognized declaration with signatories.	Provides resources.	Minimal.	Members are predominately architects and built environment professionals.	https://www.architectsdeclare.com/.
Australian Institute of Architects (AIA)	Membership organization governed by a Board of Directors, comprising elected	National with loose global reach (Founded in 1931).	MA, CP and OB: Advocacy activities include campaigning and working with	Yes, professionally orientated.	No. For Climate Action policies see: https://www.architecture.com.	Members are predominately architects and built environment	https://www.architecture.com.au/.

Table 6.1. *(Continued)*

Architectural Organization	Transparent Governance	International Reach	Protest Life Cycle	Education Programs	Presence in UNFCCC or UN-Habitat	Funding and Sponsors	Web References
	members as well as Independent Directors. Works with private/ corporate partners.		allied professionals, and all levels of government.		au/about/ carbonneutral.	professionals. Private sponsorship and advertising.	
Council on Tall Buildings and Urban Habitat (CTUBH)	Nonprofit member network organisation organised into chapters and regions. Governance: Board of Trustees; Advisory Group: Members and Staff, Multi-Disciplinary Topical Assemblies. For further breakdown See: https:// www.ctbuh.org/ about/leadership.	Global (Founded in USA, 1969 has offices in Chicago, Shanghai and Venice)	MA	Yes, professionally orientated.	No.	Members are professionals connected to the built environment.	https://www.ctbuh.org/.
Insulate Britain (UK)	Environmental activist group.	National. Initially	MA, CP, OB, CON. Advocacy		No.	Donations.	https://www.insulate britain.com/.

		associated with the global orientated, extinction rebellion.	and direct action include civil resistances including letter writing to government and civil disobedience. Targeted protest in September 2021 about insulating UK houses, specifically social housing.	Resources aimed at public.			
International Union of Architects (UIA)	Federation of national organizations. Includes professional organizations and architects in 115 countries and territories. Member Sections are divided into global regions. Partners with international and nongovernmental organizations.	Global (Founded in Lausanne, Switzerland, 1948).	MA, CP, and OB: Organizes Commissions and Work Programmes. Commissions including Architectural Education Commission and Sustainable Development Goals (SGDG) Commission.	Provides resources and charters with UNESCO for architectural education.	Yes, partnered with UN-Habitat and UNESCO. Presence at COP26.	Member Organisations.	https://www.uia-architectes.org/webApi/en/.
Open Architecture	US-based nonprofit (501c3)	Global	MA, CP, and OB: Develops		No.	Funding partners include	https://openarchcollab.org/.

Table 6.1. *(Continued)*

Architectural Organization	Transparent Governance	International Reach	Protest Life Cycle	Education Programs	Presence in UNFCCC or UN-Habitat	Funding and Sponsors	Web References
Collaborative (Related to the Former Architects for Humanity (USA- Based)	supported by a global chapter network. Leadership includes Board of Directors and Advisors.		educational programming for designers and architects. Also develops placemaking programs to promote "ownership and civic engagement on traditionally marginalized communities" (Open Architecture Collaborative, n.d).	Yes, professionally orientated.		Curry Stone Foundation, Art Works; Walter & Elise HAAS Fund as well as in-kind supporters. Also funded by donations.	
Planners for Climate Action (P4CA)	P4CA is placed under the UNFCCC's Marrakesh Partnership of Non-State actors for Global Climate Action. It is registered under	Global (Founded at the 23rd Conference of Parties (COP-23) to the UN Framework Convention on	MA, CP, and OB: Key function and activities of initiative: Implementation, Political dialogue; Technical dialogue; Goal setting (ex	Yes, professionally orientated.	Under organizational lead of UN Habitat.	No private funding.	https://www.planners4 climateaction.org/.

the NAZCA (Non-State Actors Zone for Climate Action) Platform. Lead organization UN-Habitat. 18 research and education organisations participating: "Members comprise international, regional and national associations and networks of urban and territorial planners. Individuals members are also welcoming to support the mission and objectives of the cooperative initiative. Other supporting organizations consist of research institutions and, universities, private	Climate Change (UNFCCC), in Bonn, 2017).	ante); Policy planning and recommendation; Knowledge dissemination and exchange; Knowledge production and innovation.

Table 6.1. *(Continued)*

Architectural Organization	Transparent Governance	International Reach	Protest Life Cycle	Education Programs	Presence in UNFCCC or UN-Habitat	Funding and Sponsors	Web References
	sector and other non-governmental organizations" (P4CA, n.d).						
Royal Institute of British Architects (RIBA) (UK)	Membership organization registered as a charity in the United Kingdom. Leadership includes President, Executive Team, Board, and Committees.	Nationally orientated but membership open to professionals outside the United Kingdom.	MA, CP, and OB: Advocacy activities include campaigning and lobbying the UK Government and Parliament.	Yes, professionally orientated.	Yes. UNFCCC. Presence and program of events at COP26. Signatory to United Nations Global Compact (UNGC).	Members are predominately architects and built environment professionals. Public donations.	https://www.architecture.com/.
The Architecture Lobby (T-A-L) (USA)	Two nonprofit organizations, both incorporated in the state of Connecticut: The Architecture Lobby, Inc: a 501(c)6 funded through membership dues and The	National (USA) with loose global connections.	MA, CP, OB, and CON: Advocacy and protest activities include unionizing campaigns and solidarity statements.	Yes, professionally orientated.	No. T-A-L support Statement on the Green New Deal: http://architecture-lobby.org/project/t-a-l-statement-on-the-green-new-deal/.	Members predominately architects and built environment professionals.	http://architecture-lobby.org/.

	Architecture Initiative, Inc: a 501(c)3, financed through tax-exempt donations, and the sales of merchandise, publications, etc. This arm of the organization has a Board of Directors, of which some overlap with The Architecture Lobby's BoD. (TAL, n.d.a).

Note: Protest Stage (Lakey) Life Cycle: Membership Aggregator (MA) Cultural Preparation, (CP) Organisation Building (OB) Confrontation (Con) Mass Non-cooperation (MN-c) Parallel Institutions (PI) Not Evident (NE).
Source: See https://uia.org/ database for others related agencies and organizations.

for architects and for some at least some kind of turning point. If there is any doubt that the shadows of the 1960s fall into the present, we need only look at the canon's theory bearers (Hays, 2000). But perhaps the subsumption in architectural discourse, of post-68 critical theory, has disregarded design activism and advocacy. In the present, given the existential crisis of the climate emergency, how have architects developed different modes of political activism to address this? But perhaps it is too much to expect architects to be on the streets? However, what are the characteristics of their advocacy? To ask a blunt question: Why are architects only barely present in civil society, scarcely engaged with UN-Habitat and the New Urban Agenda, or the other systems of advocacy and governance intent on enabling a transition to a carbon negative world?

The translation of the physical science into advocacy is the critical point pursued here. As suggested in previous chapters, if the city is the primary arena for climate activism, then architects have a critical role to play; a role to play in translating the physical science into design activism. This is because architecture is one of the few professions to understand the various histories of cities in terms of infrastructures, subscales, and spatial intricacies.

Architects and Protest

The activist practices of Extinction Rebellion (XR) are a benchmark for the current status of climate advocacy among architects. XR is a climate change protest collective and was formed in England in October 2018. As an organization, XR's formation, strategies, and tactics are worth placing next to the design activism of architects. XR has, of course, attracted a number of architects as its members (Dunton, 2021). However, there is no global or national section within it entirely devoted to architects. XR's first international protest named "International Rebellion" was in April 2019, and 30 XR groups quickly established themselves in different countries. Its founder Roger Hallam intended the group to be a nonviolent civil disobedience group and its first achievement was instrumental in the UK government declaring a climate emergency (Smiles & Edwards, 2021).

Notions of regenerative culture and self-care are important elements in XR's modus operandi. Self-care in XR and well-being is seen as being central in order for activists to avoid burnout. Notions of self-care are built into the group's protocols and employed as required during group and team meetings; openly expressing emotions is considered to be a systemic practicee within XR, and this extends to expressing "despair and grief, which is best illustrated by XR's emotional debrief sessions and climate grief workshops" (Westwell & Bunting, 2020, p. 547). In their fieldwork study of XR Manchester Westwell and Bunting (2020) note that this approach is intended to build within the group "a culture of care and compassion that systematically produces and regenerates its activism" (p. 548). In the XR Manchester fieldwork study, it was observed that despite these regenerative aspirations, the experiences of those members within XR related to racism were not openly discussed within the collective. Nonetheless, it was observed that there was a burgeoning discussion in XR regarding the need for conversations that "consciously embody an alternative relational politics that confronts systemic inequality" (p. 549). Moreover, Westwell and Bunting (2020) found that planet care was the area that was least developed by the XR Manchester group.

> In meetings and conversations with Manchester activists, an alternative future was rarely discussed, and when asked directly about how planet care would manifest on a societal scale, there were few ideas about what this would look like or how it would come to fruition. (p. 549)

They found that the notion of planet care in Manchester XR was an intangible and ambiguous set of aspirations; a concept not easily graspable beyond individual notions of self-care. These findings prompt a comparison to architects where intangible notions, and feel-good vibes, of sustainability are arguably never realistically defined and yet often expressed visually as alternative climate imaginaries.

In a case study of XR Norwich, Smiles and Edwards (2021) employ framing theory to understand how XR members create and frame issues of climate justice. Framing theory presupposes that sociopolitical actors or movements employ specific frames of reference and rhetoric to validate and promote particular issues (p. 3). Framing theory involves groups engaging in "collective action framing" in order to focus attention through events, situations, and social facts to convey meaning: "transforming how an object of attention is seen or understood as relating to another object or actor" (Smiles & Edwards, 2021, p. 3). In line with Benford and Snow (2000) and Smiles and Edwards (2021) state that:

> ...framing can be divided into three tasks: diagnostic framing (identifying a problem and attributing it), prognostic framing (proposing a solution to identified problems), and motivational framing (generating a rationale for acting). (p. 4)

Moreover, they additionally identify three further processes of framing:

> ...frame construction and maintenance: discursive (acts of writing and speech), contested (changes made to a frame due to its contestation), and strategic (deliberative efforts to link frames to mobilise new supporters). (p. 4)

In this detailed and nested architecture, strategic framing processes are further distinguished and:

> ...can be broken further into frame bridging (tying two separate frames together), frame amplification (the invigoration of existing values or beliefs), frame extension (incorporating new views into a

> frame), and frame transformation (when frames are transformed radically to reflect new views). (p. 4)

In their study of XR Norwich, interviews were undertaken with XR member participants to identify the different frames that they each employed. Subsequently, it was concluded that the members of XR Norwich aligned with frames of mainstream climate action rather than a more radical framing of climate justice (p. 3). Climate justice is a different, less localized framing and depoliticization of climate change and aligns to issues of international equity in the global system. These authors identify the climate justice frame emerging just prior to COP15 in Copenhagen and this frame is linked to:

> Underlying causes of climate change are linked to structural issues such as: globalisation, trade liberalisation, international debt, militarism, and global capitalism. Elites and industrialised nations are responsible for these. (Page 5, Table 1)

However, Smiles and Edwards (2021) conclude that participants in the study framed climate change in an apolitical way, avoiding thinking about it in relation to structural inequities in the global system: Again, a parallel that is also discernible in architectural advocacy and practice.

One architectural group inspired by XR and formed not long afterward was Architects Declare. In early 2019, UK architects Steve Tompkins and Michael Pawlyn privately discussed frustration with the "lack of urgency within the construction sector around the climate and biodiversity crises" and stressed how architecture needed to "find its voice" (Architects Declare, n.d.a; Hurst, 2019, p. 3). Subsequently, as the discussion continued among their wider network, a meeting took place between a number of renowned architects and practices, and all participants agreed with the idea of a joint statement of principles. The final 11 principles were initially published on the Architects Declare website from London on May 19, 2019:

- This is an emergency and radical changes are needed in the way the built environment is conceived, designed, and delivered; we seek to deliver what is necessary rather than what is perceived to be possible.
- As professional practices, our signatories have the ability to be influential across all aspects of the built environment. Together we have a strong voice.
- We are a supportive network; we do not name and shame but strive to support signatories to meet the Declaration.

  We see our principal activities as:
- Creating a devolved network for architectural practices to share information, tools, and best practice, thereby supporting one another to deliver the Declaration.

- Given the urgency and scientific basis behind the emergency, encouraging signatories to use their influence on projects and across the industry to deliver the Declaration.
- Encouraging signatories to speak out in a professional capacity, championing the radical changes needed to deliver the Declaration, whether collectively or as individual practices.
- Organizing national and international events, seminars, and conversations, to share knowledge, raise awareness, and explore how the Declaration can best be delivered.
- Working nationally and internationally with Construction Declares signatories and other like-minded organisations, to coordinate activities and be more effective.
- Engaging with the trade press and media, to support them in constructively championing the Declaration, and exposing obstacles to best practice.
- Increasing the number of practices in the network: both in Architects Declare and in the wider community that is Construction Declares.
- Using the strength of the network to engage with national and regional governments, professional institutes, and other organisations, to bring about the paradigm shift that is needed to deliver the Declaration (Architects Declare, n.d.b; Hurst, 2019).

The initial 17 signatories to Architects Declare had all been recipients of the RIBA Stirling Prize. Within the next month, over 400 UK architectural practices had committed to Architects Declare; many of them were also highly regarded and awarded within the architecture profession (Hurst, 2019). A year after its inception Architects Declare had around 900 signatories in the United Kingdom and had groups in over 20 countries around the world (Waite et al., 2020). From the beginning, the group had been the subject of sustained interest and criticism, particularly in the English architectural press from the *Architects Journal* (AJ). For example, architect Mark Elton recognized Architects Declare as a "laudable promise" while acknowledging architects may have difficulty "to transform their mindsets, become experts in building physics and engage far better with those in power, such as developers" (Hurst, 2019, p. 3). Architecture journalist and critic Richard Waite Waite (2019) questioned if anyone who had declared should be designing airports. In June 2020, *AJ* surveyed UK architects regarding the impact of Architects Declare: 50% of surveyed architects felt that it was too early to tell, 22% thought it had not been a success, and 23% thought it had been (Waite et al., 2020). However, only 11% of nonsignatory architects (4% Agree, 7% Strongly Agree) felt that the declaration had any impact on the profession or wider construction industry (p. 11).

These above debates played out in the *Architects Journal* and corresponded with various controversies regarding some of the signatories to the original declaration. By December 2020, ZHA Architects had pulled out of Architects Declare, declaiming that the declaration was "setting the profession up for failure" (Hopkirk, 2020a). This followed the withdrawal of Norman Foster regarding the sustainability issues connected to designing airports (Hopkirk, 2020a). Foster was reported as saying that innovation rather than protest was the answer to issues related to carbon mitigation (Hopkirk, 2020a). In *Building Design*, Foster was reported as saying:

> …emissions from the production of hamburgers, steaks and similar agricultural products was equivalent to those produced by air and motorised transport and called for a sense of proportion.
>
> (Hopkirk, 2020a)

In response to ZHA's walk-out Hopkirk (2020b) reported Architects Declare as saying:

> The reason we felt compelled to respond to Patrik Schumacher's recent statements was because they appeared to represent a shift away from this shared vision and thereby undermine the principles of the declaration.
>
> (Hopkirk, 2020b)

Immediately following these events, Grimshaw architects were awarded the commission for Delhi airport. In contrast to Foster's false equivalence of the hamburgers and Schumacher's histrionics, Grimshaw partner Andrew Thomas stated:

> We have no intention of leaving Architects Declare as we'd rather be part of a group which is trying to effect change than to stand outside it. We also do not think our work in the aviation sector is incompatible with AD's aims, though we are aware of the need for rapid technological and policy transformation to make aviation truly sustainable. At Grimshaw we want to be part of this solution.
>
> (Hopkirk, 2020c)

Two months after the inception of Architects Declare in the United Kingdom, architects in Australia – but not the Australian Federal Government – became the third country to declare a "climate and biodiversity emergency" with 30 high-profile architects and practices becoming founding signatories (Australian Architects Declare, n.d). Australian Architects Declare (AAD) duplicated the 11 principles of the UK declaration, but importantly it also recognized Aboriginal and Torres Strait Islander peoples' historical and holistic knowledge of country. Architect Tone Wheeler, writing in the online magazine *Architecture and Design*, offered a critical and insightful analysis of AAD:

> Few of the 30 'foundation signatories' of the declaration, outside the organisers, have a track record of sustainable design or a

> public profile of activism on climate change. This leads to suspicions of vanity virtue-signaling by signing up.
>
> (Wheeler, 2019)

While affirming AAD's advocacy, he asked if Architects do declare, revealing their personal, professional, and political position, what happens after that? (Wheeler, 2019).

In comparison to Architects Declare, the Architects Climate Action Network (ACAN) group has a different and seemingly more activist approach to the environmental politics of the climate emergency. ACAN is a network of architects, including other built environment professionals, who seek to act to "address the twin crises of climate and ecological breakdown" (ACAN, n.d.a). ACAN claims to be explicitly devoted to political action through campaigning, lobbying, and direct action (ACAN, n.d.a). Unlike other groups ACAN believes that "We are in a state of climate and ecological emergency" (ACAN, n.d.a). The group has three interlinked strategic aims: "Decarbonize Now" – rapid decarbonization, "Ecological Regeneration" – with a focus on communities and prioritizing ecosystems, and finally the "Cultural Transformation" of the architectural profession (ACAN, n.d.a).

With these essential but seemingly lofty goals in mind – for architects at least – ACAN organizes itself through a number of working groups which are described as: circular economy, education, embodied carbon, existing buildings, carbon literacy to "increase understanding of the science behind the climate emergency, planning policy, natural materials, professional standards and where the wild things aren't." This latter group works to recognize the biodiversity of the nonhuman, or wild things, with the intention of raising awareness "of the impacts of the architectural profession and construction industry on nature" (ACAN, n.d.b).

As well as a blog, the group also has an extensive presence through WhatsApp, including its Embodied Carbon group. The Embodied Carbon group won the 2021 Alliance for Sustainable Building Products (ASBP) Award, an award that normally goes to a built project (ACAN Blog, 2021a). ACAN has also developed an extensive education program, including talks and master classes, on materials, retrofitting, advocacy, and a series on the circular economy and the eight stages of the RIBA PLAN of Work (ACAN, n.d.c). On a number of occasions, the group has joined with Architects Declare. For example, the group joined with Architects Declare to urge the UK government to adopt the demands of the Insulate Britain Campaign (ACAN Blog, 2021b).

The ACAN response to the withdrawals from Architects Declare indicates ACAN's willingness to adopt some form of activism beyond simple declarations. The group wrote to Foster and Partners in July 2020 regarding Foster's involvement in the "Amaala Resort Airport and aviation expansion generally." As a result, ACAN received no response, so posted this letter to them many more times as paper aeroplanes (ACAN Blog, 2020). While this instance of activism may seem trivial, following Foster's withdrawal ACAN declared that:

> Foster + Partners have made it clear that continuing to enable aviation expansion is more important to them than being part of a collective industry effort to address the largest crisis of our time. They have signalled very clearly that tackling the climate crisis is not their priority, especially when doing so would conflict with their business model.
>
> (ACAN Blog, 2020)

The above statement also followed ZHA's withdrawal and ACAN summarized the situation in the following terms (Fig. 6.1):

> Instead of practices taking defensive stances or singing their own praises, this moment calls for an honest reappraisal of why, collectively as an industry, we have failed to address the climate crisis.
>
> (ACAN Blog, 2020)

Fig. 6.1. Insulate Britain Blocks Parliament Square, London, UK, November 2021. Image courstey of Insulate Britian. *Source*: Insulate Britain. (n.d.). *Breaking: Insulate Britain blocks parliament square*. https://www.insulatebritain.com/insulate-britain-press.

The Architecture Lobby

The activist tactics of XR, ACAN, or groups like Insulate Britain are not evident among American architects. The foremost activist group in America is The Architecture Lobby (TAL). TAL, or the "Lobby" as it is known. The Lobby is a group that promotes debates questioning labor conditions for architects and the architect's role in American society. The Lobby promotes both the value of architectural work and the value of architecture in society. Unlike other architectural membership organizations the Lobby claims to actively seek structural change in the profession. Thus, in comparison to the American Institute of Architects (AIA), the group is both more radical and more orientated to design activism. The Lobby is registered as a nonprofit organization and is governed by a board of directors elected by Lobby members with financial standing. The Lobby is loosely structured around a web of chapters, each located in a different American city. Each chapter is led by a steward in a rotating position. The chapters comprise architects in different stages of their careers, and the Lobby is also open to other allied professions. The Lobby is a global endeavor with small chapters in London and also in Melbourne, Australia. At the national and global level, the Lobby has a central board of directors of nine people as well as support from a national organiser, content coordinator, and a financial coordinator (TAL, n.d.a).

Given the lack of unionization among American architects, in many respects, TAL serves as a default or proxy union. The Lobby's 10-point Manifesto begins by proclaiming that architects are precarious workers:

We are precarious workers. These are our demands:

(1) Enforce labor laws that prohibit unpaid internships, unpaid overtime; refuse unpaid competitions.
(2) Reject fees based on percentage of construction or hourly fees and instead calculate value based on the money we save our clients, or gain them.
(3) Stop peddling a product – buildings – and focus on the unique value architects help realize through spatial services.
(4) Enforce wage transparency across the discipline.
(5) Establish a union for architects, designers, academics, and interns in architecture and design.
(6) Demystify the architect as solo creative genius; no honors for architects who don't acknowledge their staff.
(7) Licensure upon completion of degree.
(8) Change professional architecture organisations to advocate for the living conditions of architects.
(9) Support research about professional labor rights in architecture.
(10) Implement democratic alternatives to the free market system of development (TAL, n.d.a).

Interestingly – or damningly, if you want to think about it like that – the above manifesto does not address issues regarding the climate emergency. Hence, the

Lobby must then connect climate concerns to issues related to labor and capital. One way the Lobby has done this is through an internal group focused on The Green New Deal. Prior to the COVID-19 pandemic this group organized a series of events around Earth Day in April 2020 (TAL, 2020a). This included panel and speaking events focused on the economics of the Green New Deal – but no direct action. Notably, there was a Lobby Think-In in New York City, in a public discussion "on architecture's relationship to climate, frontline communities, the construction industry, labour and education" (TAL, 2020a). The Architecture Lobby in Melbourne organized an event to "reflect on the Earth" and how architects might employ their values regarding the Earth in their work. In London the Lobby bought together a panel including Anna Lisa McSweeney of the ACAN and Stephanie Edwards cofounder of Urban Symbiotics, a multidisciplinary design practice.

The link from architectural labor practices to the climate emergency is made through declarations such as "There can be no sustainable world without sustainable labor practices" and that the Green New Deal's call for a "just transition" was paramount (TAL, n.d.b.). Thus, in its work on the Green New Deal, as proposed in the American 116th Congress House Resolution 109, the Lobby developed an extensive argument for the decarbonization of the American economy (TAL, n.d.b; TAL, 2020b). Central to this argument was the assertion that the labor structure of the architectural profession must change. In summary, this polemic sought to – as set out in more detail below – reimagine and "redefine resilience," pursue "decarbonization as a social justice issue," and advocate for adaptive reuse. The polemic argued for the need to make "resiliency plans" to mitigate climate disruptions (e.g., sea-level rise) at the local community and regional scale. In more detail, the Lobby suggested that architects:

> Embrace decarbonisation as a social justice issue that calls for a reconstruction of our way of life. We must redefine sustainability to acknowledge the economic, social, racial and class-based dimensions of the climate crisis.
>
> (TAL, n.d.b)

For the Lobby, a focus on labor relations and reassessing relations between clients and architects was paramount:

> Refuse work for any client whose goals and methods – contracts, financing strategies, and labour practices – do not support a transformative redistribution of power.
>
> (TAL, n.d.b)

For the Lobby, technical innovation is seen as the main means to climate mitigation. Thankfully, in doing so, the Lobby recognized that technology was not a neutral instrument:

> …architects cannot rely exclusively on technological solutions. We must reshape our understanding of technology and innovation as design tools, acknowledging the complex power structures inherent in their development and application.
>
> (TAL, n.d.b)

Moreover, the Lobby highlighted a focus on resilience through a

> Focus on neighbourhood and regional resiliency plans in anticipation of future sea-level rise and climate driven displacement, as well as new ways of reducing emissions through the organisation of buildings, transport and infrastructure.
>
> (TAL, n.d.b)

From the above it is evident that the Lobby's position, in response to the Green New Deal, was to argue that climate change can be best mitigated by addressing the structural inequities in the American economy.

The Lobby's 2020 ABC Summer School reflects the Lobby's above approach toward climate and its attempts to read solutions to the climate emergency thorough the American capitalocene. The school entitled Architecture Beyond Capitalism was an online summer school during the northern summer of 2021. The three themes of the school were Capitalism, Labour, and Collectives. The aspiration of the school was to offer an alternative to what was missing in American architectural education. As with the response to the Green New Deal, TAL's focus was on capitalism, and its structures, in order to

> Debate the intersection of capitalism with racism, decolonisation, climate change justice and the participation of architecture in those intersections.
>
> (TAL, 2021a)

Countering conservative notions of architectural education, The Lobby formulated three demarcated themes for the summer school, capitalism, labor, and collectives, none of which directly addressed the climate emergency. The Lobby's apparent emphasis on Marxist critique overlooks the political ecology of the climate emergency in favor of an emphasis on political economy and the linkage of capitalism and colonialism. At the Capitalism session, the speaker's clearly articulated the imbrication of capitalism and colonialism (Ana María León, Andrew Herscher, Raj Patel, Huda Tayob, and Matthew Soules). Certainly, the TAL ABC Summer School curricula saw these twin evils as determining the contexts through which architecture is produced (TAL, 2021a). For example, Raj Patel, a research professor at the University of Texas Austin, examined chicken nugget production lines and how this was enabled by gendered and racialized workplaces. For Patel these lines point to seemingly disposable lives that are predominantly supported by both women and people of color. He concludes this is modern slavery akin to the transatlantic slave trade (TAL,

2021b, 40:40). Notably, in this session, Ana María León and Andrew Herscher presented from the Chicago based Settler Colonial City project; the remit of this project is to explore and reclaim the histories of Indigenous Peoples because modern architecture has not included indigeneity in its canonical narratives (Settler Colonial City Project, 2021). In the L-A-B-O-R session, Aska Welford, and Kirti Durelle of United Voices of the World (2021), presented their polemic representing the UK United Voices of the World Section for Architectural Workers (UVW-SAW). This is a union for architectural workers. What is unique about UVW-SAW is that, it open to all people who may work in an architectural office, including students, admin workers, freelance contractors, receptionists, academics, and sole traders. Membership is open to people with different qualifications and visa status who do not have the power to "hire and fire" (UVW, 2021).

The Summer school's curricula suggested an expansion of prevailing notions of architectural theory and normative practice. But, in so earnestly uncloaking the invisibilities of the American capitalocene, some things remain invisible and the climate emergency is one of these invisibilities. This concealment was despite the inclusion of diverse voices, and new perspectives on architectural theory and the prevailing canon of architectural history. The summer school indicates the way that Marxist theory and criticism is prevalent and, in some senses dominates, American architectural pedagogy. This hegemony of Marxist theory, with its emphasis on capital and labor, appears to elide the direct theorization and complicity of architects with ecocide and the climate emergency. Missing from this discourse is more recent developments in Marxist studies, such as Heron's (2021) recovery of Marxist metabolic rift theory and the work of Andreas Malm (Heron, 2021). Certainly, Guattari's project of ecosophy, of new subjectivities, of resingularization, as well as the feminist theories is missing in the account of the Lobby's efforts to prise open the hegemonic canon of American architecture.

The AIA's approach to sustainability transitions and the Green New Deal is certainly less polemical than those of TAL. Nonetheless, the AIA also declared its support for the Green New Deal in February of 2019 (AIA, 2019). The AIA's advocacy in the area of sustainability has been channelled through its Committee on the Environment (COTE) (AIA Knowledge Net, 2020). The priorities of COTE's advocacy were reinvestment in energy efficient infrastructure, enhanced climate preparation, and the decarbonization of both construction and operational building processes (Davis, October 22, 2021). Aware of the lag time between construction of buildings and energy consumption later in the life of buildings, this committee has since 2006 had an established relationship with the not-for-profit organization Architecture 2030 founded by Edward Mazria, the AIA's 2021 Gold Medalist in 2002. Architecture 2030 is a separate nonprofit research organization with the initial aim of getting to net zero emissions by 2030 in the built environment and limiting global warming to 1.5 °C. Through Mazria's efforts, the AIA adopted the challenge in 2006, suggesting a degree of foresight on the part of the AIA (Columbia GSAPP, 2021). As a part of this program AIA firms share data on energy uses in all their projects each year. The 2030 challenge has as its primary focus operational energy consumption. The tactics of this strategy is to adopt key energy targets, in addition to providing a framework of metrics, and a comprehensive data tool that allows signatories to the commitment to track progress toward net zero carbon by 2030. The specific details of the challenge and its targets can be summarized as follows:

> All new buildings, developments and major renovations shall be designed to meet a fossil fuel, GHG-emitting, energy consumption performance standard of 70% below the regional (or country) average/median for that building type.
>
> (Architecture, 2030, n.d.b)

And

> The fossil fuel reduction standard for all new buildings and major renovations shall be increased to 80% in 2020 and 90% in 2025. Carbon-neutral in 2030 (using no fossil fuel GHG emitting energy to operate).
>
> (Architecture, 2030, n.d.b)

Since its inception Architecture 2030 has also incorporated embodied carbon into its methodologies through what it designates as the principles of reuse, reduce, and sequester (Architecture, 2030. n.d.a). Despite the presumed ambiguities in measuring the success of Architecture 2030 among AIA members through COTE, it nonetheless has an interesting and impressive list of industry partners. These industry partners include a range of sustainability orientated consultants, construction and material supply companies including: Building Green, Kingspan, Sherwin-Williams, GAF Roofing, Kawneer, Lucifer Lighting Company, ROCKWOOL, Autodesk, EPIC Metals, HKS, Thornton Tomasetti, Skanska, TLC Engineering, and Sierra Pacific Windows (AIA Knowledge Net, 2020).

In the UK RIBA has put in place a sophisticated organisational infrastructure for its architect members regarding sustainability and carbon mitigation efforts. Rather than one signature initiative, RIBA's sustainability transition infrastructure is based on the RIBA Plan of Work, the 2030 Climate Challenge (not to be confused with Architecture 2030) and what it describes as an "An outcomes-based design approach" (RIBA, 2019). This approach is described as bridging the gap between "design intents and in-use performance" through the use of a number of metrics. It is hoped that the climate challenge reinforces the "feedback loop" between design briefing and project outcomes (p. 5). Of significance is the fact that RIBA has also embraced and linked its carbon mitigation efforts to the UN Sustainable Development Goals, having signed the UN Global Compact in 2016 (RIBA, 2021, p. 1:32). Consequently, RIBA developed more specific and defined guidelines regarding embodied and operational "Whole Life" carbon. In addition, RIBA has also adopted the LETI (The London Energy Transformation Initiative) targets as part of its 2030 Climate Challenge (LETI, n.d). RIBA argues that:

> ...those projects that are in the design phase at the moment should at least be targeting the 2025 targets so that we don't carry on our business as usual and totally blow that trajectory. [With] And have the construction industry contributing more C02.
>
> (RIBA, 2021,10:51)

To affect all the above, RIBA has embedded these targets and associated processes into the phases and core processes of its Plan of Work. In order to facilitate this integration, RIBA has developed a suite of guides to build capacity in architectural practices (RIBA, 2021, 2:35).

Immediately Prior to COP26 RIBA organized The Built Environment Summit linking this to the launch of RIBA's publication "The Built for the Environment Report." The report was made in conjunction with the Architects Declare UK. The report maintained faith in the ability of architects as well as the construction sector to effect ambitious change and such change was still "possible" to avoid catastrophic climate change (RIBA & Architects Declare UK, 2021, p. 11). The report propounded the ambitions of RIBA, UK Architects Declare, and the many endorsing architects, organisations, and allied business through "Ten principles for a transition to a fair and sustainable built environment" (p. 11). These are reproduced below and should be contrasted because of their specificity, with the polemics of The Architecture Lobby and the Green New Deal:

> 1. The built environment is a system, 2. Governments must deliver public resources and spending based on social and environmental justice, 3. Environmental targets must be science-based and fair. 4. Achieving net-zero requires absolute emissions reductions, not only reductions per square metre of 5. Carbon is not the only environmental indicator. 6. Mitigating and adapting to the climate and biodiversity emergency must happen together. 7. Those of us operating within the built environment must change the way we work. 8. We have the tools, knowledge, and technology to address the climate and biodiversity emergency. 9. Information should be shared openly and widely 10. Social justice must be at the heart of all action on the climate and biodiversity emergency.
>
> (RIBA & Architects Declare UK, 2021, p. 14)

The RIBA Built Environment report included a Built Environment Actor Map. The RIBA map neatly summarizes at a macro-level both the actors, policy, and regulatory frameworks that shape decarbonizing interactions across the built environment (RIBA & Architects Declare UK, 2021, p. 64). The RIBA report and map describes a global system of architects and allied multilateral organisations including: C40 Cities, Commonwealth Association of Architects (CAA), The Italian National Council of Architects, Planners, Landscape Architects and Conservators (CNAPPC), Global Alliance for Buildings and

Construction, The Royal Incorporation of Architects in Scotland (RIAS), International Union of Architects (UIA), UK Built Environment Advisory Group (UKBEAG), World Green Building Council (WGBC), and World Business Council (WBC) for Sustainable Development (p. 3). However, the focus of Table 6.1, which includes some but not all of these entities, is to assess to what degree each organization can be located on our adapted model of Lakey's protest stages; as well ascertaining to what degree each organization is embedded or active in the UN's global system. To Lakey's framework, a stage has been added in order to highlight those organizations devoted to only aggregating and building memberships: This is the Membership Aggregator (MA) stage (Table 6.1).

In contrast to all of the above organizations, the Council for Tall Buildings and Urban Habitat (CTBUH) indicates the complexities, trade-offs, and moral dilemmas when architects are involved in multilateral organizations that claim to pursue the ideal of sustainable cities. The CTBUH did not have a presence at COP26, although many of its associated groups and members may have. As noted elsewhere, CTBUH is a not-for-profit organization that describes itself as the "world's leading resource for professionals focused on the inception, design, construction, and operation of tall buildings and future cities" (Raisbeck, 2019, p. 73). CTBUH has headquarters in Chicago, and its CTBUH's supporting contributors, patrons, and donors include Autodesk, Dassault Systems, The Dow Chemical Company, Siemens, Arup, Aurecon, and Bjarke Ingels Group (p. 73). Hence, CTBUH is deeply embedded in circuits of global corporate finance, and both its research and associated rhetoric is a "mix of global city urbanism and tower technologies hyperbole" (Raisbeck, 2019, p. 73). For the most part, these efforts appear to reclaim the high-rise tower – despite literal evidence to the contrary – as a sustainable typology, and that through continuing innovation, the tower typology will not accelerate the degradation of the planet. In the tower speak of CTBUH, tall buildings are now rebranded as sustainable vertical urbanism.

For example, like the adherents of unproven carbon capture technologies, the CTBUH research journal is a curious mix of tower boosterism, incremental solutions, and technology hyperbole. In Issue 3 of the 2021 CTBUH Journal published quarterly, we are witness to the full gamut of efforts to resuscitate an unsustainable typology. For example, in a paper entitled "Is a Zero-Carbon Concrete Skyscraper Possible?", Law (2021) asserts that "We can continue to use concrete to build tall, but we have to be much smarter about the materials we use" (Law, 2021, p. 28). A future climate hopefulness is exemplified in Applegath, Veres, and Wu (2021), who discuss a hybrid timber prototype floor system (HTFS) for supertall towers. These authors state that such a system:

> …leverages energy efficiency, smart technologies, façade-integrated photovoltaics (PV), and a district energy cogeneration plant combined with an algae bioreactor, to achieve zero operational carbon on-site. Excess energy is

> provided to the surrounding neighborhood in a "net(work)-zero" carbon strategy, and a vision for future zero-carbon embodied energy is also outlined.
>
> (Applegath et al., 2021)

As might be expected, carbon optimization strategies are also canvassed. Amiri and Sakeena (2021) present a case study on, how the optimization of various structural concrete elements of towers, could contribute toward significant carbon savings. This is done by optimizing the design of core walls, slabs, and raft foundations (Amiri & Sakeena, 2021).

In 2018, Antony Wood CTBUH's CEO gave a lecture entitled "Vertical Cities of the Future" (CTBUH, 2018). Wood makes the argument that cities cannot keep expanding horizontally, and the response to this should be tall towers because with increasing density towers would be financially viable. Wood explores the idea that the tower typology might form the essential component of new cities in new "resilient" terrains when existing coastal cities have failed because of the climate emergency (CTBUH, 2018, 18:40). Wood's final point in this "utopian" excursion is to ask:

> So the final point that I'd like to leave you with is what if we had to design tall buildings or any buildings forever, forever?
>
> (CTBUH, 2018, 23:38)

Wood's pattern of rhetoric, with its Ted Talk like format, is all too familiar – and not dissimilar to the biodigital architects of Chapter 2. This form of speech can be characterized by its reductive diagrams and aphorisms, future-perfect orientation alongside big picture questions, and utopian thinking. All of this pitter-patter is designed to give CTBUH's neoliberal audience, totally embedded in development capitalism, a few trivial morsels of divergent ideas, and inconsequential future technologies, creating a smoke screen as the high-rise tower typology destroys the planet.

Another category of organizations involved in design activism is architectural entities that seem to exist in the space between corporate think tanks and strategy consulting firms. Two well-known examples of these types of firms include Forensic Architecture and the research organization of OMA architects AMO. Forensic Architecture is a research group founded by Eyal Weizman and based at Goldsmiths University of London. The group employs spatial and architectural analysis techniques to examine and explicate human rights violations (Forensic Architecture, n.d). AMO is variously described as the creative think tank of the Office for Metropolitan Architecture or as Rem Koolhaas's research organization. AMO has been involved in various exhibition designs, such as OMA's countryside exhibition at the Guggenheim and work for PRADA (Guggenheim Museum, 2020).

Territorial Agency, compared to Forensic Architecture and AMO, is a similar entity and one that is more dedicated to explicating and articulating the dynamics of climate change. Territorial Agency has a pedigree validated by its association with the Architectural Association. It is an independent organization founded by partners John Palmesino & Ann-So Ronnskog, in its own words:

> Territorial Agency is an independent organisation that innovatively promotes and works for sustainable territorial transformations. Territorial Agency is engaged to strengthen the capacity of local and international communities in comprehensive spatial transformation management. Territorial Agency's projects channel available spatial resources towards the development of their full potential.
>
> (Territorial Agency, n.d.a)

Thus Territorial Agency appears to offer a new style of architectural and urban collaboration, and a consideration of its theories are best approached through a consideration of its projects and the policies of its founders. However, where Forensic Architecture displays and presents granular images of violence and its ecologies, Territorial Agency is more concerned with global images. Territorial agency's project, the Museum of Oil, was developed in 2015 in collaboration with Greenpeace and produced with the ZKM Center for Arts and Media in Germany and was exhibited both at the Chicago Architecture Biennale and the Istanbul Design Biennale. The project targets both Shell and Exxon-Mobil as the targets of its activism. Like the work of Neil Brenner and Michael Batty, Territorial Agency employs top-down data analysis conducted and communicated through a traditional and canonical view of architectural history.

The Museum of Oil, as one of its most seminal projects, and Territorial Agency states:

> Oil is power – energetic and political: A vector of diffused and accumulated power, with leverage in both the day-to-day activities of the world-economy and in the political circuits and institutions of many national states.
>
> (Territorial Agency, n.d.b)

and:

> The Museum of Oil collects information, items, objects, data, about the different ways through which we have come closer to the decisions (it will have to be many) to keep oil in the ground, halting its combustive so-called development sprees. It portrays in detail the forms of the territories shaped by oil, indicating the main elements for the complex negotiations to disentangle ourselves from it.
>
> (Territorial Agency, n.d.b)

All of this is illustrated with seductive images described as:

> A series of large-scale Earth Observation images outline territories wrought by the oil industry: the boreal forests of Canada cut through for oil prospecting, the European Arctic with oil spills both on and offshore, the shale fracking boom in the USA, West Africa and the depletion of the Niger Delta, the militarisation of the Middle Eastern supply chain.
>
> (Territorial Agency, n.d.b)

Territorial Agency's notion of disentanglement needs to be questioned. Similar to Neil Brenner's exquisite graphs and images of Planetary Urbanisation, one must ask if the images of Territorial Agency merely document entanglement; turning the exquisite GIS images and remote sensing technology into an image-politics of fetishised and exotic images. It can be contended that Territorial Agency's spatial diagramming is all too limited to the realm of climate imaginaries.

Another group related to architects is Planners for Climate Action. This is a group aligned with UN-Habitat and was instigated at COP23 in 2017 and emerged out of the International Society of City and Regional Planners (ISOCARP) (Climate Initiatives Platform, n.d). The group seeks to build knowledge and skills capacity for young planners. The group has affiliates across the globe, but most affiliates are in the Asia Pacific region. For many planners dedicated to protecting both built and cultural heritage, climate change is of real concern. In UN-Habitat the group has highlighted its work in built and cultural heritage contexts. But this also extends to work the group has done, for example, through the Canadian Institute of Planners with the Nunavit Adaptation Plans. For the Planners for Climate Action group, heritage is seen as a key driver to build capacity in order to account for climate risks. The preservation of cultural heritage, through planning instruments, is seen as a way to regenerate and revitalize urban areas. This group sees urban planning as having an important role in capacity building through heritage regeneration in order to meet the increasing effects of climate change. In this approach urban heritage systems are seen as being key drivers of decarbonization, climate adaption, and what they designate as "nature-based solutions." Such as green roofs, green infrastructure to enable cities to be "more resilient" (Planners for Climate Action, n.d; UN-Habitat, 2020).

While some architects might be cynical about the sentiments of planners in regard to heritage revitalization, there is another more important point to be made: There is no parallel organization representing architects in the UN system. As nonstate actors, the various climate initiatives that architectural entities do promote are not part of or captured as information in the UN system. This point is that few architectural organisations are embedded in or engaged with the UN's global organisations and system of governance concerning climate change. As indicated in Table 6.1, the attendance of architectural groups at the COPs or the World Urban Forum is often ad hoc and more opportunistic than strategic (Fig. 6.2).

Fig. 6.2. XRWA Grandparents Lock-on Protest at Woodside Headquarters in Perth, Australia, December 2021. Image Miles Tweedie Photography and Extentiction Rebellion Australia. *Source*: Extinction Rebellion. (n.d.). https://ausrebellion.earth/news/xrwa-grandparents-lock-on-protest-at-woodside-hq-in-perth/.

Conclusion

The spectrum of activism evident in Table 6.1 indicates that architects prefer indirect action over direct action. For the architect, environmental politics – with few exceptions – is something that is effected through their existing professional organisations. The "Taxonomy of Action" set out by Jensen, McBay, and Keith (2011) in their book *Deep Green Resistance* indicates a broad range of "Acts of Omission" and "Acts of Commission" that encompass both "Indirect and Direct Action" (p. 243). Acts of omission include strikes and walk-outs, and other forms of noncooperation. While acts of commission include direct actions such as obstruction and occupation, indirect actions include lobbying and protests and symbolic acts. With this taxonomy in mind, it can be seen that few architectural organizations pursue radical acts of either indirect or direct climate action. To reiterate, at one end of the spectrum, direct action, though civil disobedience, is not something that the architects of the Global North have readily adopted – ACAN or architects in XR or Insulate Britain are the exception. In the middle of this spectrum, we are witness to groups, such as Architects Declare, effecting mitigation action within the boundaries of architectural firms. RIBA – in contrast to say the American AIA – has established an ongoing response that includes interventions in professional education, linking its own Plan of Work to the SDGs and direct political advocacy, in the UN-Habitat and COP system, alongside general public advocacy. And at the other end of the spectrum, it can be seen how groups such as CTBUH employ media advocacy and "research" paradigms, not unlike the biodigital architects of Chapter 3, to disguise any real action, political or otherwise, in relation to the climate emergency.

The willingness of architects to pursue either nonviolent civil disobedience or violent civil disobedience is a point that needs to be broached. As seen in this and previous chapters, architectural advocacy against the climate emergency and ecocide is in parts riven by a mainstream culture of make-believe – of climate imaginaries – a self-serving culture that the American environmentalist Derek Jensen would characterize as bright green lies. Certainly, groups like ACAN, and arguably RIBA, have done much to become involved in direct action through various alliances. However, overall the range of architectural action has been indirect and limited: it has not included acts of omission, and for the most part, it is only through indirect action that architectural groups have acted in any organized way. At best, lobbying, symbolic acts, and education awareness and training have been the main modes of architectural activism in the climate space. Capacity building through education to enable architects to better measure and ascertain carbon emissions has been a key feature of more effective approaches. Nonetheless, Architects, by and large, appear to eschew any design activism and direct action that might, for example, damage the reputations of their clients in order to protect their own revenue flows. Architects will still work for fossil fuel companies – or build carbon loaded infrastructure – and at the same time declare their good intentions in other sustainability venues and avenues.

The real question for architects is not simply what they need to do and what new agencies they need to develop to keep global warming to 1.5 °C with deep 2030 emissions reductions – beyond the rhetoric of net-zero framings. In architecture there is a spectrum of political action that exists between the pantomimes of Holcim and CTUBH, the middling manifestos of TAL and the Green New Deal, the activist pedagogies of Architects Declare and ACAN then – at the other extreme – the spatial practices and actions of Insulate Britain and XR. Given this spectrum, the real question for architects is to what degree should they adopt either civil disobedience and in what form should this disobedience take. Architects have never had a debate to discuss the possibilities of either violent or nonviolent protest in order to save the planet. To what degree should architects adopt new forms of design activism and resistance in order to avert the ecological destruction of the planet? As Andreas Malm has argued: when is it time to blow up the pipelines? (Malm, 2021). Such a debate within the architectural profession will arguably, reframe, broaden, and give clarity to the current mainstream debates and efforts – as documented in this book – which are absolutely mired in incremental business as usual habits. Both architectural theorists and practitioners subscribe to the idea that architecture's autonomy affords it a curious separation from capitalism. This engenders a false criticality that is simultaneously critical of capitalism and yet seemingly uncoupled from capitalism. This paradoxical illusion – generated through a disassociation from the contingencies of architectural practice – forces the following question: how does this type of criticality, and its attendant spatial practices, enable the dismantling of the infrastructures of ecocide. Architects, who pursue criticality within the canonical tradition devoid of signifiers (Eisenman), or a tradition overloaded with signifying ironies of the canon (Stirling), can no longer say that their gestures in any way produce new subjectivities that will avoid climate catastrophe. Where is the voice of the Earth in the dance between these binary positions, between an architecture empty of politics and the other a political cacophony of concepts and symbols? Certainly, the Earth's voice is better heard in the turn to affect, to situated bodies, to notions of care, to slow spatial practices, or the discovery of new liberatory practices within the environmental humanities (Frichot, 2020). But are all of the things – that seek to disrupt the autonomy of the prevailing canon – no matter how desirable too complacent? To pursue this line of thought to its logical conclusion: given the evolving arena of environmental politics, at what point will architects consider the need for direct action that will sabotage carbonized infrastructure?

References

ACE (Architects Council of Europe). (2014). The role of the Architectural Profession In Delivering Responsible Design. Retrieved from https://www.ace-cae.eu/uploads/tx_jidocumentsview/ACE_URBA_ISSUES_2014_EN_FORMAT_A5_HORIZ_PANORAMIQUE_WEB_01.pdf

American Friends Service Committee (AFSC). (2012, March 1). George Lakey on strategy for a living revolution [video]. *YouTube*. Retrieved from https://www.youtube.com/watch?v=Af5DPmP6Pc

American Institute of Architects (AIA). (2019, February 8). AIA supports green new deal framework. Retrieved from https://www.aia.org/press-releases/6105450-aia-supports-green-new-deal-framework-

American Institute of Architects (AIA) Knowledge Net. (2020). Committee on the environment. Retrieved from https://network.aia.org/communities/community-home?CommunityKey=3b790506-aca5-4eff-aaf6-8a7b553dc0ef

Amiri, A., & Sakeena, N. (2021). Towards the carbon-neutral high-rise: The role of embodied carbon. *CTBUH Journal*, 3, 36–43. Retrieved from https://global.ctbuh.org/resources/papers/download/4495-towards-the-carbon-neutral-high-rise-the-role-of-embodied-carbon.pdf

Applegath, C., Veres, C., & Wu, T. (2021). The zero-carbon hybrid future of tall timber. *CTBUH Journal*, 3, 20–27. Retrieved from https://global.ctbuh.org/resources/papers/download/4493-the-zero-carbon-hybrid-future-of-tall-timber.pdf

Architects Can (ACAN) Blog. (2020, December 4). ACAN response to architects declare withdrawals. Retrieved from https://www.architectscan.org/post/response-to-architects-declare-withdrawals

Architects Can (ACAN) Blog. (2021a, March 11). ACAN embodied carbon Campaign wins ASBP 2021 award. Retrieved from https://www.architectscan.org/post/acan-embodied-carbon-campaign-wins-asbp-2021-award

Architects Can (ACAN) Blog. (2021b, September 14). ACAN's statement on the aims of insulate Britain. Retrieved from https://www.architectscan.org/post/acan-insulate-britain-statement

Architects Can (ACAN). (n.d.a). About. Retrieved from https://www.architectscan.org/about

Architects Can (ACAN). (n.d.b). Action: Working groups. Retrieved from https://www.architectscan.org/action

Architects Can (ACAN). (n.d.c). Events: Upcoming events. Retrieved from https://www.architectscan.org/events

Architects Declare. (n.d.a). About us. Retrieved from https://www.architectsdeclare.com/about

Architects Declare. (n.d.b). AD activities and principles: Background information. Retrieved from https://www.architectsdeclare.com/resources#architects-declare-resources

Architecture 2030. (n.d.a). Embodied-carbon-actions. Retrieved from https://architecture2030.org/embodied-carbon-actions/

Architecture 2030. (n.d.b). The 2030 challenge. Retrieved from https://architecture2030.org/2030_challenges/2030-challenge/

Australian Architects Declare. (n.d.). Australian architects declare climate & biodiversity emergency. Retrieved from https://au.architectsdeclare.com/

Benford, R. D., & Snow, D. A. (2000). Framing processes and social movements: An overview and assessment. *Annual Review of Sociology*, *26*(1), 611–639.

Climate Initiatives Platform. (n.d). Planners for climate action (P4CA). *Climate Initiatives Platform/UN Environment Programme/UNEP DTU Partnership*. Retrieved from https://climateinitiativesplatform.org/index.php/Planners_for_Climate_Action_(P4CA)

Columbia GSAPP. (2021, January 30). Edward Mazria [video]. *YouTube*. Retrieved from https://www.youtube.com/watch?v=rI6wErOFF4Q

Council on Tall Buildings and Urban Habitat (CTBUH). (2018, April 28). CTBUH 2017 Australia conference-Antony Wood "vertical cities of the future" [video]. *Youtube*. Retrieved from https://www.youtube.com/watch?v=aP-lp45Mpc0

Davis, R. M. (2021, October 22). Advocacy update: Staying optimistic. *American Institute of Architects (AIA) Knowledge Net*. Retrieved from https://network.aia.org/blogs/michael-davis/2021/10/22/advocacy-update?CommunityKey=3b790506-aca5-4eff-aaf6-8a7b553dc0ef

Dunton, J. (2021, August 26). Architects join extinction rebellion demonstrations. *The Architects' Journal*. Retrieved from https://www.architectsjournal.co.uk

Forensic Architecture. (n.d). About. Retrieved from https://forensic-architecture.org/about/agency

Frichot, H. (2020). Infrastructural affects: Challenging the autonomy of architecture. In M. Jobst & H. Frichot (Eds.), *Architectural affects after Deleuze and Guattari* (pp. 10–25). London: Routledge.

Guggenheim Museum. (2020, January 29). See countryside, the future at the Guggenheim [video]. *YouTube*. Retrieved from https://www.youtube.com/watch?v=eTbH5RWb66o

Guillemot, H. (2017). The necessary and inaccessible 1.5 C objective. In S. C. Aykut, J. Foyer, & E. Morena (Eds.), *Globalising the climate: COP21 and the climatisation of global debates* (pp. 39–56). Abingdon: Taylor & Francis.

Hays, K. M. (Ed.). (2000). *Architecture theory since 1968*. Cambridge, MA: MIT Press.

Heron, K. (2021). Dialectical materialisms, metabolic rifts and the climate crisis: A Lacanian/Hegelian perspective. *Science & Society*, *85*(4), 501–526.

Hopkirk, E. (2020a, December 4). ZHA pulls out of climate pact accusing architects declare of 'setting profession up for failure'. *Building Design*. Retrieved from https://www.bdonline.co.uk/news/zha-pulls-out-of-climate-pact-accusing-architects-declare-of-setting-profession-up-for-failure/5109364.article

Hopkirk, E. (2020b, December 9). Architects declare offers olive branch to ZHA and fosters after walk-outs. *Building Design*. Retrieved from https://www.bdonline.co.uk/news/architects-declare-offers-olive-branch-to-zha-and-fosters-after-walk-outs/5109453.article

Hopkirk, E. (2020c, December 10). Grimshaw lands Delhi airport job-but vows to stay in architects declare. *Building Design*. Retrieved from https://www.bdonline.co.uk/news/grimshaw-lands-delhi-airport-job-but-vows-to-stay-in-architects-declare/5109498.article

Hurst, W. (2019). Architects declare: Architecture firms are rallying to the climate change cause with an 11-point action plan. But have they understood the transformation in practice this will commit them to? *The Architects' Journal*, *246*(11), 4–6.

Jensen, D., McBay, A., & Keith, L. (2011). *Deep green resistance: Strategy to save the planet*. New York, NY: Seven Stories Press.

Lakey, G. (1973). *Strategy for a living revolution*. New York, NY: Grossman.

Law, D. (2021). Is a zero-carbon concrete skyscraper possible? *CTBUH Journal*, 3, 28–34. Retrieved from https://global.ctbuh.org/resources/papers/download/4494-is-a-zero-carbon-concrete-skyscraper-possible.pdf

LETI. (n.d). About. Retrieved from https://www.leti.london/about

Malm, A. (2021). *How to blow up a pipeline*. London: Verso Books.

Open Architecture Collaborative. (n.d). [website]. Retrieved from https://openarchcollab.org/

Planners for Climate Action—P4CA. (n.d). Planners for climate action. Retrieved from https://www.planners4climateaction.org/

Raisbeck, P. (2019). *Architecture as a global system: Scavengers, Tribes, Warlords and Megafirms*. Bingley: Emerald Publishing Limited.

Royal Institute of British Architects (RIBA). (2019). *RIBA sustainable outcomes guide*. Retrieved from https://www.architecture.com/knowledge-and-resources/resources-landing-page/sustainable-outcomes-guide

Royal Institute of British Architects (RIBA). (2021, July 22). RIBA 2030 climate challenge: Introduction for architects [video]. *YouTube*. Retrieved from https://www.youtube.com/watch?v=1qpCVGwEmvY

Royal Institute of British Architects (RIBA) & Architects Declare UK. (2021). *Built for the environment Addressing the climate and biodiversity emergency with a fair and sustainable built environment*. Retrieved from https://www.architecture.com/knowledge-and-resources/resources-landing-page/built-for-the-environment-report

Settler Colonial City Project. (2021). SCCP at urgent pedagogies. Retrieved from https://settlercolonialcityproject.org/SCCP-at-Urgent-Pedagogies

Smiles, T., & Edwards, G. A. (2021). How does extinction rebellion engage with climate justice? A case study of XR Norwich. *Local Environment*, *26*(12), 1445–1460. doi:10.1080/13549839.2021.1974367

Territorial Agency. (n.d.a). Territorial agency. Retrieved from https://www.territorialagency.com/

Territorial Agency. (n.d.b). Museum of oil. Retrieved from https://www.territorialagency.com/museumofoil

The Architecture Lobby (TAL). (2020a). T-A-L Earth week. Retrieved from http://architecture-lobby.org/event/t-a-l-earth-week-2020/

The Architecture Lobby (TAL). (2020b). T-A-L statement on the green new deal [google document]. Retrieved from https://drive.google.com/file/d/1Sa1zlFpwEVLY96LM8cqkOAIETF9RlC2m/view

The Architecture Lobby (TAL). (2021a). About: Architecture beyond capitalism school sumer school. Retrieved from https://abc.architecture-lobby.org/

The Architecture Lobby (TAL). (2021b, June 20). C-A-P-I-T-A-L-I-S-M—ABC school—speaker session [video]. *YouTube*. Retrieved from https://www.youtube.com/watch?v=FElmIgvS1Vk&t=3706s

The Architecture Lobby (TAL). (n.d.a). About: Manifesto. Retrieved from http://architecture-lobby.org/

The Architecture Lobby (TAL). (n.d.b). T-A-L statement on the green new deal. Retrieved from http://architecture-lobby.org/project/t-a-l-statement-on-the-green-new-deal/

UN-Habitat. (2020). *The tenth session of world urban forum, cities of opportunities: Connecting culture and innovation. Abu Dhabi, report, September, 2020 [report]*. United Nations Human Settlements Programme (UN-Habitat). Retrieved from https://unhabitat.org/report-of-the-tenth-session-of-the-world-urban-forum

United Nations Climate Change. (2021). *COP26 Glasgow climate pact [report]*. Retrieved from https://ukcop26.org/wp-content/uploads/2021/11/COP26-Presidency-Outcomes-The-Climate-Pact.pdf

United Voices of The World. (2021). Sectors—Section of Architectural Workers (SAW). Retrieved from https://www.uvwunion.org.uk/en/sectors/architectural-workers/

Waite, R. (2019, November 16). Climate crisis: Should architects still be designing airports? *The Architects' Journal*. Retrieved from https://www.architectsjournal.co.uk

Waite, R., Ing, W., & Hurst, W. (2020, June 18). What has architects declare achieved? *The Architects' Journal*. Retrieved from https://www.architectsjournal.co.uk/news/what-has-architects-declare-achieved

Westwell, E., & Bunting, J. (2020). The regenerative culture of extinction rebellion: Self-care, people care, planet care. *Environmental Politics*, *29*(3), 546–551.

Wheeler, T. (2019). Tone on Tuesday: On architects declaring. *Architectural Design*. Retrieved from https://www.architectureanddesign.com.au/people/tone-on-tuesday-on-architects-declaring#:~:text=Everyone%20is%20declaring%20a%20climate%20emergency.&text=This%20is%20to%20be%20especially,percent%20in%20construction%20and%20materials

Chapter 7

Towards a Political Ecology of Architecture

> The supreme remedy and insurance against any dangers lies in the magic of mist, called *Kayag'u*, which, side by side with Kula magic, and the magic of canoes, is the third of the indispensable magical equipment of a sailor.
>
> (Malinowski, 1922/2002)

Introduction

For all their dreams, regimes, and conceptual sleights, architects cannot hear the voice of the Earth as it reels from the ecological violence of humanity. What else can be said about the above landscape of vaporous theories, biophysical alchemy, indirect political action, and green sleights of hand? How else might one begin to apprehend all of the preceding as a political ecology? In mapping this ecology, even partially, we return to Bogman Palmjaguar and rather than seeing Bogman as the outsider, we can place him at the center of things: "trained conservationist and outlawry of the rocks" (Bogman, n.d). Rather than seeing Bogman as one to be taken into the autopoietic culture of architects, to be given succor on a stormy night, gifted a token legitimacy, we can better place Bogman at the center of things. It is then the architect who is the schizo: It is the architect on the edge of the forest, architects as both sorcerers and shamans, distanced observers, on the edge of ecological catastrophe; seemingly content with their own company, their only pact with each other, and the violent sounds of the dying Earth are barely heard in their own chants and misereres.

In his anthropological study of the Kula exchange system of the Trobriand Islands, Malinowski describes how "the realm of the magical" is described by the word megwa:

> In native terminology, the realm of the magical is called by the word megwa, which describes the 'magical performance', the 'spell', the 'force' or 'virtue' of magic, and can be used as adjective to describe in general everything which presents a magical character.
>
> (Malinowski, 1922/2002, p. 330)

Architects, Sustainability and the Climate Emergency, 189–213

Published under exclusive licence by Emerald Publishing Limited
doi:10.1108/978-1-80382-291-420221007

Design is a word that – like the word *megwa* in the Trobriand Islands – is used by architects as both substantive and adjective (Malinowski, 1922/2002, p. 330). In architecture this word describes performances, rituals, conceptual spells, always reinforcing its own force and virtues. Design, like *megwa*, can also "be used as an adjective to describe in general everything which presents a magical character" (Malinowski, 1922/2002, p. 330). For architects design is a type of magic – and for many architects it is the magic. As Hagen (2017) notes in her study of Snöhetta: "Crafting magic concerns the ritualistic practices of securing the dialogue between materiality and the virtual or imaginary" (p. 24). Pursuing sustainability through design, for many architects, is a practice of alchemy, a spatial practice linked to the desire to create new substances via the transformation of processes of construction and existing materials. Through these spatial practices architects create a dialogue between the materialities of all that they touch and the sustainable and climate imaginaries of the future.

The magic of architects is produced and reproduced through a professional culture of autopoiesis. No other than the architect Patrick Schumaker has argued that architecture is an autopoietic system: "most adequately grasped if it is analysed as an autonomous network (autopoietic system) of communications" (Schumacher, 2011, pp. 62–69). Schumaker limits their definition solely to the material artifacts of architectural practice and so excluding any entanglement with socio-natures. An autopoietic system is a closed system that determines and sets its own organizational rules, creating its own boundaries, and its structures and internal dynamics are self-generated. As Maturana and Varela (1980) explain, autopoietic systems are determined by their structures, and changes in structure are determined by an overall unity – even if the trigger for change is exerted from the external environment. The overall unity of the system determines how triggers a responded to. Autopoiesis suggests professions can be understood as social ecologies and need not be regarded as bounded groups of actors "engaged in collecting, organising, and operating upon 'facts' from the environment" (Ahuja, Nikolova, & Clegg, 2017; King, 1993, p. 218; Raisbeck, 2020).

Autopoiesis emerged as a concept in the biological sciences and has a legacy in systems theory and cybernetics (Varela, Maturana, & Uribe, 1974). While autopoiesis has its origins in the biological sciences, Luhman reformulated the theory for the social sciences (Luhman, 1987; Maturana & Varela, 1980). Luhman (1987) posited that autopoiesis is a theory that views communication as the primary component of social systems – and not simply the communication of material artifacts cut off from all socio-material entities. In the social sciences, autopoiesis has been explored in various settings. Notably, Åkerstrøm and Stenner (2019) employ Luhman's theories to identify discourses of management centered on "potentialising technologies." In other words, technologies related to the mantras like "the future of the future." If autopoiesis is a broad framework that describes the workings of the architectural profession, then the responses of architects to the climate emergency exist within a self-generating and self-perpetuating cultural system. The practices of magic evident in the political ecology of architecture take place within this autopoietic culture; and this magic is the dynamic means by which architects structure and self-generate their responses to the climate emergency.

Magical Scripts

Magic Scripts

There are a number scripts and codes from which architects draw the power of design magic. These scripts are assemblages of theories and broader ideas rather than project-specific spatial practices or construction innovations. These scripts often align with several well-worn tropes evident in architectural discourse. Some of these magic scripts emanate from various overdetermined binary conceptions or different *Weltanschauungen*. Consequently, in attempting to transform the gap between various binaries (such as technology and nature), a more reductive or limiting idea is the result; often found alongside a simplistic and didactic hyperbole. Architects often appear to fall into the traps of critical responses to Cartesian dualisms identified by Val Plumwood (1993, 1995), as discussed in Chapter 2. As also indicated in previous chapters, binary scripts and codes are drawn upon by architects to develop charismatic rhetorical spells or performances. For architects, these spells often point to processes, either imagined or actual, of binary reconciliation. For example, stories that seek to harmonize humans and nature or culture and technology. In their book *Planning Wild Cities*, Steele (2020) articulates some of the sensibility of these binary scripts:

> The anthropocentric city is founded on ontological binaries (humans/nonhumans; organic/artificial; living/non-living; wild/tame), underpinned by a metaphoric spatiality of centres, distances and boundaries. At the centre of this ontological space is the sovereign human subject.
>
> (Steele, Wiesel, & Maller, 2019, p. 414)

By drawing on the hyperbole of the binary, both architects and their allied urban policymakers easily create and craft green city narratives. These narratives enable the production of harmonious images of sustainability, but this mode of production does little to articulate the intricacies of sustainability transitions in cities. In fact, these dualistic scripts may have the opposite effect in limiting architectural agency in relation to the climate emergency.

To put the argument a little more precisely, and following Nightingale et al. (2019), for architects the parsing and overdetermination of "identifiable and external climate threats" is rooted in a structural binary; a binary that disassociates climate and environmental factors from social factors; even as architects simultaneously appear to seek to reconcile these complex factors:

> At the heart of this dualism is the notion that we can separate out the 'climatic' drivers of change from 'social' ones, therein creating a set of policies to deal with climate change – adaptation and

> mitigation – in isolation from the broader trajectories of socio-environmental change.
>
> (Nightingale et al., 2019, p. 2)

In enumerating Nightingale et al. (2019) in the architectural context, as well as the loss of architectural agency through ineffective narratives, there is another problem: conceiving of climate action where "social and environmental processes" are ideated in design thinking "as separate yet interacting forces" leading to "a set of largely technical solutions for policy and practice" (p. 2). As defined by Nightingale et al. (2019), technical solutions can be defined as follows:

> By technical, we mean responses that impart new infrastructures or that seek to create biophysical changes; examples which vary in emphasis and across scales from water control infrastructures such as irrigation, drinking water and dams, to tree plantations, or geoengineering for carbon sequestration. (p. 2)

As articulated in previous chapters, the technical narratives of architects reconciling the biological and the physical are produced in various ways. Firstly, through the biophysical imaginaries of unbuilt and seemingly utopian projects – as noted in Chapter 3. Secondly, in built projects where technical solutions are evident in various design and construction innovations, although these innovations are no more than isolated prototypes – as noted in Chapter 5. The technical narratives of architects anticipate a future-perfect Anthropocene in climatic equilibrium, arguably an erroneous, if not dangerous, misreading of the physical science of IPCC's reports. These technical narratives conceal others ways of architectural knowing and echo the perspective of Nightingale et al. (2019) regarding policy:

> While the vocabulary of impacts, adaptations, emissions and mitigation has become firmly embedded in the policy sphere, we believe this framework conceals ways of knowing that might help more effectively address the climate predicament and foster determined action. (p. 2)

For the architects the technical solution conceals ways of knowing that arise from other nontechnical and spatial domains, including the "lived experiences, cultural memories, and the arts" (Nightingale et al., 2019, p. 2). Many architects will, of course, say that these nontechnical domains are at the center of their discipline; despite ignoring the crowding out of the architectural humanities by technical research and so-called science in architecture schools – admittedly an assertion not fully pursued here. As a result, sustainability transitions are only seen in technical terms, and effective climate action appears to be stuck. Building little technical machines and prototypes on the edge of the woods is more performance and entertainment, drawing our gaze away from ecocide, than an engagement with the vicissitudes of the climate emergency. Technical and policy solutions related to biophilia, bio-inspired design, regenerative design, and robotic manufacturing disguise or render invisible co-emergent climate morbidities that are never thoroughly examined by architects.

Alongside the above, one of the most potent magical scripts in architecture is the systems script. This overarching script is drawn on to make sustainability spells and magic seem more potent; any urban entity can be described as a system or part of a system: individual projects, circulation routes, villages, cities, and other urban entities as systems. For architects, words like ecologies and ecology – also the study of interdependent systems – are conflated and confused with the systems script. Perhaps this is what Grose (2017) means when they argue that designers need to "shift thinking away from ecological opinion without facts" (p. 7). Lilienfeld's (1978) early account of systems theory still resonates, and this account relates how the term "system" moved into new territories – such as architecture. The term became

> ...evermore loosely defined and impressionistic in usage, sometimes being used in Bertalanffy's sense of an open system, sometimes in the cybernetic-communications theory sense, sometimes (e.g. in operations research) to mean simulation models, and often in a blend of all these. (p. 227)

For the architect, of course, the systems script has a distinct genealogy in architectural discourse. The various technological systems that emerged out of the military-industrial complex during the Cold War were often celebrated by 1960s architects and allied polemicists. For example, John McHale (1967), reiterating the hopes of Bertalanffy's General Systems Theory, promoted a "cross disciplinary" fusing of knowledge across disciplines; suggesting cybernetics would link together "computer technology, management and systems science, social engineering and bio-engineering" (p. 64). The system approach had a concrete form in the Operations Research and systems analysis tools of the aerospace programs of the 1960s. This approach was fundamentally about a "science of administration" (Lilienfeld, 1978, p. 105). An approach by which systems analysts had "made serious claims" in fields as diverse as "transportation, waste disposal, health, education social welfare, foreign policy, crime control, traffic control, psychotherapy, population control, food supply, futurology, as well as understanding of life and the universe" (Lilienfeld, 1978, p. 104). And so why not architecture and cities? And so, the systems urbanist, nowadays as they did in the 1960s, arrogates to themselves a "universal knowledge" that may also encompass "biological, medical and social systems" (Lilienfeld, 1978, p. 104).

Like architects, the systems analyst is also a shaman: "The systems analyst claims to master and combine mixed disciplines and to exploit the gaps between the historically evolved disciplines of knowledge" (Lilienfeld, 1978, pp. 111–112). However, as Lee (2020) points out in her exploration of the RAND corporation the "research protocols and agendas, related to operations research, cybernetics, and systems discourse" can also be linked to a midcentury visuality and image politics emerging from the enmeshing of both "hard and social sciences" (Lee,

2020, p. 50). Lee (2020) posits that this approach laid the groundwork for neoliberalism and came to be embodied in the figure of what they call the aesthetic strategist – exemplified at RAND corporation by the nuclear strategist Albert Wohlstetter. Lee (2020) goes on to highlight the invasion of the visual arts by the paradigms of the Cold War laboratory. And indeed, the traces of the laboratory as an episteme can be seen in many of the visual images and rhetoric of architects. The laboratory concept, with all its associated visual images and rhetoric, enables architects to conjure up a cloak of scientific knowledge – for example, in the algae experiments' Ecologic.

More could be said about laboratory magic and its image politics in architecture. More could also be said about the historical genealogies of systems thinking in the work of figures such as Christopher Alexander, Cedric Price, Norman Foster, Richard Rogers, and urbanists such as William J Mitchell. But more urgently it is noted that the numerous sorcerers of systems thinking in the 21st century, as outlined in previous chapters, have all too easily developed the magical scripts of systems thinking. The magic of systems enables architects and urbanists to bring together biological and physical worlds as an aesthetic of concepts and images. These images denote underlying principles, structures, morphologies, and mechanics in order to bolster, rather than liberate us from – as Francois Roche suggests – the authority of the systems of control. For many architects and urbanists, system worlds abound, and the scripts of these worlds, as well as binary scripts, might also be likened to what Malinowski names as systemic magic.

> Such magic consists of a connected and consecutive body of incantations and concomitant rites, no one of which can be torn out of its sequence and performed by itself. They have to be carried out one after the other in a determined order, and the more important of them, at least, can never be omitted once the series has been started.
>
> (Malinowski, 1922/2013, p. 319)

In architecture at one level, the opposite seems to be true: systemic magic is less about sequences and determined orders and more about selecting and recombining various rites; rites selected from a pool of binary tropes and system narratives. At another level, there is always an attraction to systems of modular growth, modularity, optimized sequences, and control through ordered measurement and repetitive production. As with the polemics of many biodigital architects, there is always a railing against linearity in favor of complexity and complex systems. And yet the indicative prototypes of these systems ends in the most ordinary aesthetics of modularity and become little more than digital dashboards.

But there is also a void in these scripts. And that is the void of politics. There is no need for the work of politics or citizenship – except for those elite cadres of aesthetic strategy – because when all the different systems merge together, conflict is reconciled and smoothly recalibrated in dashboards of data. For the proponents of the systems city, both imagined and partially realized in data warehouses, all of its clever algorithms and self-organizing resilience is sufficient to mitigate the climate emergency. Without politics or citizenship, the systems city has no need to engender the liberatory practices suggested in Constant's New Babylon or the urbanism of the Situationists. It is also a city far removed from traditional authorities, such as the Kula, and their own systems of governance. Consequently, its illusory modalities of multidisciplinary coproduction are complicit in both producing climate change and the perpetuation of ecological violence.

Sustainability Magic

Numerous spells are drawn from and recombined by architects using the above scripts. Malinowski identifies and painstakingly classifies the different types of spells in a taxonomy of Kula magic. In their taxonomy, there are three essential aspects to the magic that enable magical ceremonies to be distinguished between the "formula, the rite, and the condition of the performer" (Malinowski, 2017, p. 215). Some spells are uttered either with or without rites, and these rites may also include various offerings and invocations. The condition of the performer is also important depending on the spell; for example, when performing rain magic "the magician paints himself black and has to remain unwashed and unkempt for some time" (Young, 2017, p. 220).

Sustainability Spells and Formulas

To be clear, the contemporary magic of architects is certainly less complex than what Malinowski describes. Sustainability is conjured in spells with the magical formulas of the rating tools, various rites of certification, and regimes such as the Green Building Councils, Passive House (PassiveHaus), LEED, BREAM. There are also the algorithmic formulas of parametric design: simply through their use, sustainable outcomes are seen to be achieved; All of these formulas for making spells mean that there is no need for architects to question the neoliberal city – or capitalism for that matter. These various formulas and recipes, with their appearance of quantitative certainty, encourage processes of managerial governance that rarely question the ambiguities of measurement. Moreover, these formulas engender the autopoietic ceremonies and rituals of architects through various industry awards. Even the condition of the performer, the body of the architect, puts into circulation positive green messages in various media ecologies. But there is, of course, a problem with this formulaic magic, and that is each spell's appearance of quantitative certainty is illusory. For instance, the affirmation of seemingly green buildings, such as rise towers, disguises the ambiguity inherent to the various systems of measurement.

To reiterate the above, the magic of formulas and their use as instruments of boosterism belies the real ambiguity behind these various formulas. This measurement relies on considering their top-down or bottom-up processes, the extent to which supply chain boundaries are included in the measurement, data transparency, and differences between different jurisdictions – not to mention the data that product manufacturers might provide (Crawford et al., 2021). Many architects and their clients seem oblivious of these ambiguities and nuances. Some also appear to resist the measurement of embodied carbon in buildings, and others are simply satisfied to adopt operational emissions as the only benchmark. For example, as reported in early 2021, Foster and Partners were reluctant to endorse RIBA's *Built for the Environment* report, which advocated for the need to measure both Embodied Carbon in the production of buildings as well as the operational emissions produced by buildings once they are built (Ing, 2022). Embodied carbon is thus a key and emerging issue in this arena. At the time of writing it will be critical to see how embodied carbon issues are presented in the IPCC AR6 Working Group III report including buildings. The formulas of rating tools such as LEED and BREAM might give an architectural project and its clients the imprimatur that a particular building or project is green. But alarmingly, such formulas utilized as green spells by architects have not been widely adopted in the development and construction industry (Lorch, 2017).

Sustainability Rituals

To pursue the analogy of magic further there are also the rites associated with various sustainability spells. In contrast to formulaic magic there are also more complex spells entwined with various offerings and invocations. In these spells it is the western canon of modernist architecture that is invoked and paid homage to, in order to give a particular sustainability spell more power. The rites and offerings of these spells are often performed in the lecture theaters, or through writing, in the elite architecture schools where the prevailing canon is rarely questioned, or little attempt is made to dismantle it – unlike the work of those architects discussed in Chapter 4 (Tayob, 2019; Varela et al., 1974). he green rituals that invoke the canon enable a process where theories, policies, and techniques of decarbonization can be enfolded and normalized within this prevailing and powerful canon. This phenomenon is most evident in American architecture schools as arguably this is where the didactic teaching of the western and modernist canon has the most prominence.

The machinations of how decarbonization is mistakenly theorized and enfolded into the canon are evident in the fall 2019 edition of *Log* entitled *Overcoming Carbon Form*. This title alone suggests that tectonic form cannot ever be abandoned and that the atmospheric gas carbon must also be reified in concrete form in orthodox architectural theory. In the keynote essay entitled "Architecture and the Death of Carbon Modernity," Elisa Iturbe (2019) proposes a simplistic and linear chronology of human history, claiming that "There have been three paradigms of energy capture throughout human history – foraging, agriculture, and carbon-intensive fossil fuels" (p. 11). This historical chronology is problematic from a number of different viewpoints. Both the terms used – foragers – and representation of history through a number of selected phases suggests a framing that does not appear to question colonialist notions of progress. In the desperation to conceptualize the concept of carbon form as a historical construct Iturbe (2019) writes: "The shift from a nomadic to an agricultural society gave rise to towns, villages and cities" (p. 11). And then:

> The shift from an agriculture to an industrial society driven by fossil fuels gave rise to factory towns, global trade networks, suburbs and megacities. This most recent of these spatial paradigms, which I call *carbon form*, is the focus of this issue. (p. 11)

This prompts a more detailed question: who exactly are foragers? Are the foragers, and nomadic societies, in this context Indigenous People? And does this infer that foragers and nomadic societies are somehow backward? In a footnote supporting these assertions, Iturbe cites *Foragers, Farmers and Fossil Fuels: How Human Values Evolve* (Morris & Morris, 2015). Ian Morris, a professor of classics at Stanford University, takes an evolutionary approach to history, arguing that colonization and conquest have benefited humankind. Clearly, the naive simplicity of Iturbe's history of urbanization mirrors the popularisers of world history such as Morriss and is blind to the colonizing – and racist – tropes of these historical framings.

Iturbe's (2019) approach signals a different kind of magic reliant on magical formulas. In defining the category of carbon form in spatial terms alone, they hope to "implicate architectural and urban form in the creation and unfolding of the anthropogenic climate crisis in a way that looks beyond the immediate quantification of a carbon footprint" (p. 12). Iturbe (2019) argues that the identification of carbon form, defined as the spatial expression of carbon energy, is a "site of intervention" and architecture has a "significant role in defining the outcome of this increasingly uncertain phase of human and planetary history" (p. 12). Iturbe's conceptualization and focus on carbon form retains all the elements of an approach grounded in canonical theories, the rudimentary identification of historical phases, architecture as an autonomous domain, the reification of the abstract concept of carbon form, and the delimiting of architecture to spatiality alone. More disturbingly, there is also the language. The naming of the nomadic versus the sedentary, an entire universe of terms and soundbites such as "petro-culture" and: "Architecture cannot address the causes of the climate crisis without recognising carbon modernity as a continuous cultural and material foundation for building and form" (Iturbe, 2019, p. 17) And then added to all of this the aphorism that "form follows energy" (Iturbe, 2019, p. 20). And through this ritualistic writing, homage is rapid to the pedagogies of the architectural canon, reinforcing its linear progression in time securing the dialogue between the architectural production of carbon emissions to both the virtual and imaginary histories of the western architectural canon.

Iturbe's position quite rightly calls into question the ability of techno-utopianism to transform the Anthropocene. Iturbe (2019) rejects techno-utopian innovations and what they call "avant-gardist tabula rasa" in the place of these modes they posit that decarbonization in architecture is a "theoretical question for architecture and the city" (p. 21). This is a position that abandons any consideration, either as a solution or critique, of climate technology and building systems (p. 21). Why think about the politics of building technologies and systems? Instead, Iturbe's (2019) project is to "unearth new sites of architectural agency" rejecting "technological advancements in building systems" making the call for architects "to reject unlimited growth and lay the groundwork for the coming transitions in energy and social form" (p. 21). All that is left to do is to treat decarbonization as "a cultural project and a formal problem" – and after all, what else can architecture do?

In the above delimiting of spatiality alongside the retreat to tectonic form and culture, the legacies of the American historical canon are normalized. To rephrase this more succinctly, in retreating to both form and culture, Iturbe invokes the American architect Peter Eisenman's legacy and fetishes (Iturbe coauthored a book with him. Eisenman & Iturbe, 2020). What shadows the narrative of carbon form is Esienman's rejection of technology in the 1960s and his warm embrace of syntactic language and form without politics. For both Eisenman and Iturbe, the architect must desperately cling to and proclaim a disciplinary authority; and a commitment to an autonomous discipline shielded from the consumption of a consumerist and business orientated profession. In following Eisenman and eradicating the political Iturbe appears to cling to little more than a depoliticized and impotent climate advocacy.

Central to this ritual, and its repeated invocation, is the mantra that architecture has no role to play in politics. Theory in this incarnation has become the practice of extracting and eradicating politics from architecture. Architecture's disciplinary specificity reigns supreme, and there is no room for any resubjectivation, ala Guattari, because for Eisenman:

> I believe there is a subject, I believe there is a disciplinary specificity to all disciplines and what I believe one is looking to do – in addition to anything else – is find what that disciplinary specificity is in architecture.
>
> (Archinect, 2004)

As Douglas Spencer (2016) succinctly notes in his book *The Architecture of Neo-Liberalism* this leads to a situation where:

> Architecture has sought to sanitise the 'madness of theory', to dispossess itself of its theoretical demons. Theory has been worked over until it can be put to work for and within neoliberalism. The 'smooth' and the 'folded' have been instrumentalised in the affirmation of flexibility and compliance, 'complexity' employed to dissimulate neoliberal imperatives as the laws of nature. (p. 72)

For Eisenman architecture is cut loose from politics, political agency, and activism, and politics has no place in architectural theory except as a deflective mechanism to reinforce its own boundaries. This chasm of politics and allied myopia – mired in American exceptionalism – evident in the never-questioning of the canon is, given the climate emergency, both remarkable and reckless. It is tragic that a younger generation of American theorists and architects are entranced by this type of magic.

Shamans: The Condition of the Performer

There are so many shamans and warlords in architecture: In their performances, their own condition and status as a performer is always present in the background. More often than not, as Raisbeck (2019) points out, this status is related to the pedigrees of the world's elite architecture schools. Each performance and media appearance of the shamanistic architect highlights and expresses conditions of privilege. For those of us consigned to the periphery of this global system, as a result of our own architectural education, we are either trash or invisible; spectators subject to and witnesses of the performances of the privileged architectural stars and visitors. Thankfully, in recent years there has been a concerted effort to question the conditions of these shamanistic performers through a consideration of race, ethnicity, class, and heteronormativity (Castricum, 2017). Nonetheless, the condition of the performer, through their status and the amount of symbolic capital they personify, often remains paramount.

And while all of the above may sound like a kind of denouement, what requires noting is that these performers serve a purpose and have a political base – to evoke a term associated with Trump. To maintain brevity one example will suffice: The architect Patrik Schumacher – perhaps the supreme neoliberal shaman – is linked to an extensive network connected to those who have locked themselves into the technologies of the parametric. A list of this base, the obvious inheritors of Sogo, of the Metabolists, of Archigram, of Bucky Fuller and Cedric Price, is extensive, and mostly it is centered on a kind techno-scientific – if not nutty – academicism; an academicism that has no need for Eisenman's interest in culture. These networks encompass The Association for Computer-Aided Design in Architecture (ACADIA) and the work emanating from the DRL lab at the AA. The arterial and aesthetic deployments of Tom Wiscombe in Los Angeles and Rob Stuart-Smith at the Weitzman school of design. In Australia, these networks extend to the kitsch pragmatism of Jane and Mark Burry at Swinburne University; at RMIT, the fragments and skeins in the work of Roland Snooks, Alisa Andrasek founder of Bio thing and Professor of Design Innovation and Tom Kovac. The so-called archaeologist of this digital and parametric legacy is Greg Lynn. The intention here is not to be churlish but to point to the extensive networks of those aligned with this unfortunate techno-utopianism. This techno-utopianism is paradoxically achieved through the optimized pragmatics of computing, arguably built on a largely masculinist shamanism. It will be interesting to map, as 1.5 °C seems increasingly hard to reach, and as climate impacts kick in, the slow and concomitant realization in these networks of parametric sorcery that rapid decarbonization matters.

The architectural shaman, as noted in previous chapters, can also produce magic on a global scale. This is what magic allows architects to do. Bjarkes Ingel's effort, and grand spell, to master plan the entire fucking planet places monomaniacal global strategies above the particularities of local and regional climate conditions. Ingel's approach – as Keith Critchlow was to remark of Fuller – embodies a belief in a synergetic world view, a kind of new religious science, "one of an orderly, inevitable eternal universe"; a universe in which "scientists have always found by experimentation an a priori orderliness of nature" (Critchlow, 1976, p. 296). Echoing Fuller's propensity toward snake oil, Ingel's sorcery – as well as that of the above technical fixers – adopts design thinking as a kind of religious science. As a result, all of these grand spells create a global mist that disguises the complexity of the climate emergency.

But there is also another category of shamans who are aligned with and model themselves on management consulting and think tanks. One example is Territorial Agency, a collaborative entity with a pedigree validated by its association with the Architectural Association. It is an independent organization founded by partners John Palmesino & Ann-So Ronnskog, in their own words:

> Territorial Agency is an independent organisation that innovatively promotes and works for sustainable territorial transformations. Territorial Agency is engaged to strengthen the capacity of local and international communities in comprehensive spatial transformation management. Territorial Agency's projects channel available spatial resources towards the development of their full potential.
>
> (Territorial Agency, n.d.a)

In some ways, the work of Territorial Agency is not dissimilar to Forensic Architecture (Forensic Architecture, n.d). However, where Forensic Architecture displays and presents granular images of violence, Territorial Agency is more concerned with creating diagrams and images using global information. Territorial Agency's project the Museum of Oil was developed in 2015 in collaboration with Greenpeace and produced with the ZKM Center for Arts and Media in Germany. The project was exhibited both at the Chicago Architecture Biennale and the Istanbul Design Biennale (Forensic Architecture, n.d). In the project both Shell and Exxon-Mobil are the targets of Territorial Agency's activism. Moreover, like the work of Neil Brenner and Michael Batty, Territorial Agency employs top-down data analysis conducted and communicated through a traditional and canonical view of architectural history. The Museum of Oil:

> …collects information, items, objects, data, about the different ways through which we have come closer to the decisions (it will have to be many) to keep oil in the ground, halting its combustive so-called development sprees. It portrays in detail the forms of the

> territories shaped by oil, indicating the main elements for the complex negotiations to disentangle ourselves from it.
>
> (Forensic Architecture, n.d)

All of this is illustrated with seductive images described as:

> A series of large-scale Earth Observation images outline territories wrought by the oil industry: the boreal forests of Canada cut through for oil prospecting, the European Arctic with oil spills both on and offshore, the shale fracking boom in the USA, West Africa and the depletion of the Niger Delta, the militarisation of the Middle Eastern supply chain.
>
> (Territorial Agency, n.d.b)

Territorial Agency's production of knowledge practice still needs to be questioned because it indicates a modality that establishes the voice of the shaman through a think tank model. For example, Territorial Agency in their seminar series "Plan the Planet" the consultant's shamanistic voice combines with a spatial analysis that offers an overdiagnosis related to the dilemmas of the technology, the city, and architecture. In these utterances, we see the 1950s and 1960s canon laid before us and combined together into a rhetoric of shamanistic questions: The story of architecture in the 1950s via Team X and CIAM X at Dubrovnik, the story of architecture in the 1960s via Fuller and Tyrwhitt. Mixed into this background are notions of the Holocene, the Anthropocene, contemporary disasters like Fukushima, and a seemingly scientific verbiage borrowed from the radiological sciences such as notions of half-life and half-lives (AA School of Architecture, 2019). Territorial Agency is reminiscent of Neil Brenner's exquisite graphs and diagrams related to Planetary Urbanisation (Urban Theory Lab, 2021). Arguably, this model of magic never really engages with direct political action or collective design activism; it is an image-politic of fetishized and exotic global images. In diagramming the pathologies of the Earth's ecological destruction, these images blur the line between entertainment and climate action. As Rainer Maria Rilke might describe these images, they enable us to be "witness to the world's disgrace" and little more (Dowrick, 2008).

Invisibility Magic

There are, of course, other scripts and other theoretical methodologies that appear, through their different framings, that appear to make sustainability and the climate emergency invisible. These approaches seem to have no need for spells that create the mists or pretenses of sustainability. These are magic spells, an invisibility magic, that cloaks the climate emergency so it cannot be seen at all. For example, Brott (2019) employs a psychoanalytic approach to read the iconic architecture of star architects through a psychoanalytic viewpoint of guilt (Brott, 2019). Brott (2019) posits that the iconic architecture of the global system is intimately linked to capital. Privileging Adorno over the Marxist materialism of Tafuri they contend that:

> ...iconic buildings are not a reflection of external reality, but a regression to an unconscious architecture polemic, that secretly carries out the disciplinary rituals of modernism's death, and seeks to make good on the liabilities of architectural history. (p. 41)

Subtly invoking Hiroshima and Auschwitz, Brott (2019) dissects modern architecture, and its complicity with capitalism, in psychoanalytic terms through the concepts of systemic guilt, a desire for eternal life, psychic debt, wish fulfilment and the redemption of the modernist architectural project (p. 42). For Brott (2019), "immortality has always been the end goal of technology" (p. 41), and modern architecture appears not simply in the guise of an "instrument of technical progress" but as a cult of "death resistance" (p. 41). Certainly, the iconic architectural project is a candidate for psychoanalytic readings. But is the production of redemptive and iconic images in the face of climate-induced extinction best explained through concepts of collective guilt and Adorno's notion of eternity resistance in popular culture (Brott, 2019, p. 41). While Brott argues this guilt can be located in architecture's failure to realize its own Elysium dreams, in the face of the climate emergency, the context of modernity has now irreversibly shifted. Rather than seeing iconic projects through the Freudian notion of guilt, they are best, and can also be seen, seen as the hallucinatory representations and mists of desire – and not as death resistance. If modernity and its twentieth-century terrors was a trauma for architects – as Brott suggests – the climate imaginaries of architects are now less death wish-fulfilment and more a soothing magic that creates a psychotropic deliria; green deliria that are more divertissement than the impulse to guilt-ridden redemption. If the function of architecture as a redemptive icon is redundant, then there is little else to do but to revel – and party – in the world of the climate imaginaries.

However, within architecture – outside of psychoanalysis – there is still the prevailing script of Marxist criticism. This is invisibility magic conditioned and perpetuated by the totalizing universe of the American capitalocene – perhaps best observed from a periphery. This focus on capital and architectural labor is evident in the work of Peggy Deamer, Felicity Scott, and Douglas Spencer; as a result of this immersion in a specifically American political economy, the positional differences between these theorists is minimal (Deamer, 2020; Scott, 2016; Spencer, 2016). Yet, on occasion, they are made to appear larger than what they are, as evidenced by Spencer's 2021 resignation from The Architecture Lobby. But more significantly, the need to focus on architecture's relationship to the Anthropocene, and the Earth, are cloaked by a discourse of architecture's complicity with the capitalocence. There are a few breakthrough moments in this invisibility magic: the exploration of capitalism and decolonization or the tracing of environmental genealogies in the canon, but for the most part, climate issues are largely omitted.

As a self-described critical theorist of architecture in this arena, Douglas Spencer's outlook exemplifies this invisibility magic. Spencer's earlier work in the decade was to react to, and rightly so, the betrayal of architectural theory by what Spencer described as the object turn. This terminology referred to the "Object Turn" edition of *Log* 33 published in February 2015. As Dutto (2018) notes in a commentary, Spencer via the best "dialectical-materialism orthodoxy" savaged this "object turn." A turn to the object evident in the work of philosopher Graham Harman and the object-orientated ontologies of biodigital architects such as Tom Wiscombe (Dutto, 2018). Spencer (2016) argues that the turn was a "marketing device" fetishizing the media of magazines and does nothing but confirm capitalist production, abolishing "whatever hierarchy of values between human and nonhuman objects" (p. 240). For Douglas Spencer (2016) this apparent cleansing of architectural theory in the work of biodigital architects leads to a complicity with free markets. As he succinctly notes in his book *The Architecture of Neo-Liberalism*:

> Architecture has sought to sanitise the 'madness of theory', to dispossess itself of its theoretical demons. Theory has been worked over until it can be put to work for and within neoliberalism. The 'smooth' and the 'folded' have been instrumentalised in the affirmation of flexibility and compliance, 'complexity' employed to dissimulate neoliberal imperatives as the laws of nature. The way of the market comes to appear as the way of the world, and vice versa. (p. 72)

Yet mistakenly, in prioritizing a critique of political economy, Spencer pursues the above argument in a later critique of Bruno Latour (Spencer, 2021). Despite the earlier invocation of the nonhuman as physical, they also rail against the presence of the on-human as species. For Spencer (2021) Latour's materialism and interest in environmentalism is seen as escapist:

> Current imperatives to acknowledge and respect the nonhuman speak, though, to a specifically contemporary desire to escape the supposedly anthropocentric perspective of modernity, and with it the environmental, social and even philosophical problems for which it is held responsible. (p. 150)

Going on to argue that these seemingly flat ontologies, because they do not privilege human actors, can be regarded as romantic novelties – probably because they lie outside of Marxist criticism. For Spencer (2021) the real horror is not the prospect of climate extinction and ecological destruction but the idea that capital and labor are overlooked in Latour's ontology:

> There is no room in Latour's flat ontology, either, for the agency of the 'automatic subject' that is capital; for Marx the 'all-dominating economic power of bourgeois society'.

> Overlooked in not qualifying as a thing, slipping beneath the horizon of visibility, the agency of the biggest actor of them all is rendered obscure in Latour's schema. (p. 154)

Spencer critiques a joint essay between Latour with Yaneva, (2012) in which they argue that Le Corbusier's Assembly building in Chandigarh, India, should be read not as an iconic monument but as:

> …the multifarious manifestations of recalcitrance of this building, resisting breezes, intense sunlight and the microclimate of the Himalayas, etc.
>
> (Latour & Yaneva, 2012; Spencer, 2021, p. 155)

Spencer cannot abide with these readings of the Assembly building as a life-world rather than an iconic monument. Spencer decries the loss of human agency in the capitalocene, but also an approach that abandons and reduces the primacy of capital in the Anthropocene; any approach not predicated on "exchange relations produced within capitalism" (p. 154). As they state, Latour's crime is to associate with materialist feminist philosophers like Val Plumwood (2012) and their ilk. For Spencer Latour and Yaneva's crime is to be impoverished materialists because they reject:

> All that is abstract, conceptual, subjective, theoretical, anything that cannot be grounded and brought to "earthly account," is to be excised from consideration. (p. 155)

Spencer thus throws an invisibility cloak over the necessity for exploring new subjectivities in the face of the climate crisis. Spencer, in following his spiritual mentor Manfredo Tafuri, is locked into the insularity of the American capitalocene with all of its rituals of an institutionalized Marxist critique. For Spencer (2021), there is no interest in opening up new avenues of political ecology; rather, there is a dogged clinging to the dictatorship of the political economy. Anything outside of this economic realm, such as lifeworlds, Indigenous Knowledge Systems, nonhuman species, the ecologies, and voice of the Earth do not matter – all that matters is a critique of the capitalocene.

Taboos

> The associated concept of taboo, as Malinowski observed, is covered by the Kiriwinian word bomala (with suffixed possessive pronouns). It means a 'prohibition', something which a man is not allowed to do under any circumstances. It is used for magical taboos, for prohibitions associated with rank, for restrictions in regard to food generally considered as unclean, as, for example,

> the flesh of lizards, snakes, dogs and man. There is hardly any trace of the meaning of 'sacred' attached to the word bomala.
>
> (Malinowski, 2017, p. 204)

Architectural practice theory, the theory *in and of* practice, is for many architects and architectural theorists a kind of taboo. This erasure of the contingencies and politics of architectural practice, as a subject, has increasingly come to bedevil architects who subscribe to this exclusionary outlook. Too often, the theory–practice split is often expressed as a binary between so-called avant-garde architects – who seek the validation of the canon, wallowing in a specific disciplinarity – and their frenemies, the mercantile architects. This leads to a position as Manfredo Tafuri was to state in regards to the New York Five:

> They – like Kahn and Venturi – are only emblems of a 'condition' of intellectual work; or better, of the remnants of an intellectual work that believes itself capable of constituting itself as a closed space defeated from intolerable encounters that resound with noises whose mere echo seems deafening.
>
> (Tafuri, 1987, p. 297)

The taboo of finance is one of the ways that architecture's specific disciplinarity is maintained. Architects will readily embrace critiques of political economy, if only to underscore a binary separation between those following the precepts of the canon and those architects wedded to mercantile profits. The prohibition on engagement with finance begins in the architectural schools and is promulgated through crude notions of the architect as either artist or business person (Raisbeck, 2019, pp. 71–73). Hence the prohibition regarding finance: Finance is always an unclean subject and one that can never be truly sacred as a focus of study in the historical canon. This leads to the situation where architectural practice, as a subject in architecture schools and as a field of knowledge, is seen as being secondary to design.

For those architects and theorists of political economy, operating in the enclosed spaces of the architectural canon, there is a sense of practising a jiu-jitsu like discipline, a keeping of faith with the emblems and badges of intellectual work constituted by the canon. But this binary outlook, between autonomy and the mercantile, highlights the faux subversiveness of architects who claim agency but ultimately find that they are incapable of enacting any political project. On the one hand, some architects will reject the mercantile connotations of practice and yet readily use theory to embrace its forms and technologies. Architects of the left – if we can call them that – have through meme-like concepts, slogans, and abstractions scrubbed theory clean of politics, allowing architecture to operate both autonomously and flexibly in free markets. Consequently, these architects unwittingly allow the dangerous production of carbon to continue.

The taboo of finance leads to a situation where, among other issues, the most critical issues related to climate finance and climate justice in the global sphere are missing in the architectural theorization of the climate emergency. It could be argued that these issues, because of their global scale, related as they are to the financial circuitry of the global economy, are of no relevance to architects. And yet, as described herein, architects are obsessively drawn and fascinated by the global scale – especially when it can be read through the received canon. The issues of climate justice and development and the need to provide climate finance to developing countries are absent from most architectural theories.

Spells of the Mist

> The supreme remedy and insurance against any dangers lies in the magic of mist, called kayga'u, which, side by side with Kula magic, and the magic of the canoes, is the third of the indispensable magical equipments of a sailor.
>
> (Malinowski, 1922/2002, p. 188)

and

> Now the leading idea of kayga'u is that it produces some sort of mist. The mulukwausi who follow the canoe, the sharks and live stones which lie in wait for it, the depth with all its horror, and the debris of the canoe ready to harm the owner, all these are blinded by the mist that arises in obedience to these spells.
>
> (Malinowski, 1922/2002, p. 189)

For architects, there are also the mists created by the magical scales of the planet and of the city. These mists, once invoked, protect architects from the climate emergency by allowing them not to see it. As suggested above in relation to Bjarke Ingels, Territorial Agency, Ecologic, as well as old Buckminster Fuller – and in the way architects conceptualize the Anthropocene – the invocation of the global scale as a spell creates a magic mist that affords architects protection from danger. Moreover, as evident in urban planning discourse, the question of scale is one that should not be ignored. Thus the concept of scale, in both its planetary and city incarnations, its ebbs and flows in architectural and urban theory, must be addressed in any discussion of sustainability, cities, and architecture. Debates around city scale, among urban theorists, are much more prevalent in comparison to architects.

A recent instance that summarizes and allows us to enter this debate is Ihnji Jon's (2021) effort to both revisit and reclaim the notion of scale in their examination of two city dyads: Darwin versus Tulsa and Cleveland versus Cape Town. Jon (2021) affirms that in the Anthropocene, cities should still be regarded as a central subject of urban theory and that a politics of the Earth, an "earthly politics," one that enables the mitigation of carbon, is best prosecuted through the city concept. Jon articulates the history of scale in urban theory, pointing out the problem of reconciling planetary and global conceptions of city scale and local, territorial bounded and situated notions of scale. To state Jon's (2021) case a little more expansively, they argue for the need to create a middle ground between two seemingly dichotomous poles – perhaps too easily characterized as the local and the global. Jon's project is, in their own words, to argue and propose how cities can perform "environmental politics beyond environment" (Jon, 2021, p. 1). By this Jon means the need to forge an approach to environmentalism that is neither essentialist nor about simply protecting nature (p. 2).

However, like other global city scholars, Jon argues that the cities, and the city scale, are critical in considering how to manage the planet. Jon's effort is to reconfigure existing notions of scale through functionalist and territorial framings: such as local, regional and national, urban and rural; all, in order to create new liberated "reconfiguration of scale" in urban theory. Jon's solution, somewhat predictably, is to adopt a middle ground that does not threaten the theoretical primacy of the good Anthropocene city in urban theory. It is a project that seeks to be both bottom-up and top-down. In their words, this is a search for political ecologies and framings that mediate in a middle ground across both the local and global. In this view, cities are not abandoned as a theater of politics and action but can then jump up a scale, affording an opportunity for a new "earthly politics," that advocates pro-nature and pro-environmentalist initiatives that are motivated and driven by our attachment to local territorial experiences (Jon, 2021, p. 8). Jon's (2021) position is a little too reasonable, given the pressing demands of the climate emergency, although subtly argued. Jon's position, like other urban scholars, is all about the city; with no jump down in scale or any descaling cities that might give consideration to architecture and its spatial practices in cities discourse.

In defending the city and adopting a middle ground of scale, Jon must counter – like Douglas Spencer in architectural theory – the "new ecologists" whose sources, not dissimilar or far removed from this book, can be found in the work feminist criticism and political ecology in the work of Isabelle Stengers, Donna Haraway, Karen Barad, and Anna Tsing (2021, p. 8). Jon recognizes that the contribution of these thinkers. This contribution is credited as the idea that, the binary understandings of nature and essentialist environmentalism create unnecessary boundaries between humans and nonhuman ecologies. At the same time Jon argues that a focus on local or bottom-up scale, as seemingly espoused by these new ecologists, is overly enamoured with the local and reifies a kind of "romanticism" of the local. Jon (2021) states that this "ever-narrowing" notion of scale – a framing that seems linear and telescopic – is problematic because ground-up "direct democracy" practices may engender negative consequences at larger scales (p. 17). They write:

> The idea of collectively deciding the boundaries of a territory - through discussions among the locals who believe they belong to a certain territory - remains theoretical or practically naive, since, in reality, it would mean that the scale of these territories should remain small enough for direct democracy to be operationalised. (2021, p. 17)

Jons' outlook is an instrumental subjectivity, exhibiting an outrageous arrogance toward the so-called locals; through these enactments of territory political power, the agency of ecopolitics is only granted at the level of the city – never through the spatial practices of architects. This echoes the reappearance, if not the continuation, of the hegemony of colonial territory. Nevertheless, Jon's approach points to how the idea of scale, alongside the centrality of the city itself, is a critical element in urban theory debates. Yet architects appear to be outside of these debates in urban studies, and their invocation of the city or planetary scales through corresponding signifiers via language or visual images, as argued in previous chapters, only creates a further mist of confusion.

Neil Brenner's work, in its diagrammatic and visual formats, has been seductive to those architects and urban designers wishing to create the magic of a global mist around their work (Dorignon, 2015; Urban Theory Lab, 2015). However, Brenner's work, as well as Ihnji's above, points to how removed architects are from the architects the ongoing debates and legacies (e.g., Manual Castells and Henri Lefebvre) in urban theory. Brenner, in the most recent iterations of his own writing, appears to question the planetary urbanization, city-centric, and scalar foundations of urban theory by suggesting that there is a crisis of urban studies (Brenner, 2019). This crisis is called into being by acknowledging the "hegemonic dispositif of city-centric urban knowledge"; and the "mainstream interpretive framework, urbanization is said to entail the universal diffusion of cities as the elementary units of human settlement" (Brenner, 2019, p. 317). In doing so, Brenner (2019) highlights and calls out:

> ...triumphalist discourses that celebrate the accomplishments of contemporary urban 'science' and depict the city as the optimal scale for economic growth, innovation, technopolitical management, and/or ecological modernisation. (p. 308)

Clearly, all of the above are all too evident in architectural discourse. In support of a new project of theorization and reflexive questioning in urban studies, Brenner cites the antipathy to city centrism, for example in the work of Cairns (2018). Who argues that city-centrism comprises a set of mutually reinforcing networks and actors with a shared vocabulary that leads to:

> ...foregrounding the effects of concentrating and concentrated city settlement types. The policy directions that can follow from such a city-centric understanding of urbanisation often prescribe intensification of city settlement: dense, compact, accessible and mixed-used city types. (2018, p. 304)

Unlike the systems managerialists such as Kvan and Burry, Brenner (2019) sees the need to develop new instruments:

> As the spatialities of the capitalist urban fabric restlessly mutate, so too must we continually reinvent the repertoire of concepts,

> methods, and cartographies through which its uneven patterns, variegated pathways, crisis tendencies, and contradictions are investigated. (p. 309)

For architects such investigations may only be maps of patterns that do little more than conceal the nature of the climate emergency. Having created a crisis in his field Brenner then brings methodological pluralism in as the savior: "the urgency of developing epistemically heterodox research strategies in the field of critical urban studies" and "a de facto multiplication of urban lifeworlds, knowledge formations, interpretive schema, methodological tactics, and research agendas" (Brenner, 2019, p. 330). Brenner thus rejects a science of urban life that is:

> …a singular, 'monist, authoritative' in favour of a more specialised model where embedded in the urban studies is specialised forms of knowledge production (whether from the social sciences, the spatial and environmental humanities, ecology, or environmental studies) within 'engaged pluralist,' intercultural modes. (p. 317)

Brenner's solution then is to adopt a functionalist approach that highlights, and arguably privileges, the production of expert knowledge. It begs the question: do these intercultural modes of coproduction include an engagement with the inhabitants of informal settlements, or will these people always be reduced to diagrammatic lines in Brenner's exquisite planetary maps. Nonetheless, in questioning the city – and its concomitant scalar lense – as the paramount urbanizing element, we might come to see more clearly in the mist.

In urban planning research centered on the Pacific Island states, Alexei Trundle (2020) argues that urban informality has "latent capacities" that can "enhance the climate resilience of urban systems" (Trundle, 2020, p. 1). Trundle points to what he calls "sub-city" scales and analysis, which enables a revised reading of urban resilience; a reading that might previously have been concealed in larger-scale analysis of resilience and postdisaster scholarship. Trundle's contribution to debates about informality is based on interviews with 57 households from across the Pacific Islands. This research emphasizes how informal processes enhance resilience through the "maintenance of ecosystem services, use of kinship and familial networks, and translocation of traditional knowledge" (Trundle, 2020, p. 1).

Conclusion

Brenner's, Jon's, and Trundle's work point to the artificial knowledge divisions that exist between architects, urban planners, and the dreams of a so-called urban science (Acuto et al., 2018). On the one hand, architects will readily create mists through magic that is related to both the planetary and city scales. This is done by adopting and reproducing ideas of densification, concentration, colocation, and condensed spatialities in which different functions are given presence and activated. Architects, in abandoning these scales, might see the capacity of informal settlements to shape their own theoretical and spatial subjectivities: not as entities to be theorized from macroscale perspectives or planetary subjectivities, instead to listen intently to the subcity scales and capacities emerging from this engagement. Certainly, as seen herein, architects and architectural theorists have not engaged with the global cities discourse in urban studies; but this discourse – already multiplying and fragmenting – given its grasping primacy vis à vis the city appears to have emphatically denied the granular knowledge and skein of connections that architectural knowledge might cede. So, architecture is not part of the urban debates of scale; it has a scant presence in global cities discourse, and for the most part, urban theorists know little about the possibilities of its disciplinary logics. Everything in Jon's accounts of scale, like other global cities authors, is about the lines and demarcations of territory. This is the same for Brenner, where expert knowledge is eternally demarcated between different expert fields and in Brenner's exquisite mappings where graphic pixels are vividly inscribed over the Earth and its First Nations.

Although it may seem somehow novel to abandon the primacy of the city, with a little more imagination, the urban scholar might see the potential of recognizing subcity scales encompassed by architecture. For both architects and urban scholars, this new eyesight might allow the small operationalized and direct democracy territories of First Nations people, such as those in Australia, to be seen. However at present, the magic mists of architects and the instruments of planners have no appetite for these noncity territories. Through the mists of magic architects cannot see much: the nonhuman, living mythical beings speaking through the Earth itself or collapsed notions of space and time. The dismissal of the new ecologists lacks imagination and clings too readily to traditional notions of territory: territories that are measured, gridded, bound, owned, and then colonized with histories of ecocide and genocide; all the territories of the Capitalocene.

What can be said about this landscape of sustainability enchantment, magic, and mists? The political ecology of architecture is a realm of superstition with its own occult histories and talismanic diagrams. A territory around which there are numerous esoteric cults of theory and greenwashed image practices. It is through these image practices that architects turn for comfort. All of these spells of theory and practice enable architects not to see the depths of the ecological crisis, not to see ecocide or hear the voice of the Earth. The voices and signals emanating from ecological systems are barely heard within architectural discourse. It is only in the mists created by the magic of climate imaginaries unhinged from the realism of climate science or the contingencies of decolonization and development that architects feel safe. As the real mists of the climate emergency approach, as methane emissions accelerate, and ice sheets melt, architects appear content to observe and dream on the edge of the forest.

References

AA School of Architecture. (2019, October 19). Climate summit: Plan the planet-John Palmesino [video]. YouTube. Retrieved from https://www.youtube.com/watch?v=Szo0zjnMNQs

Acuto, M., Parnell, S., & Seto, K. C. (2018). Building a global urban science. *Nature Sustainability*, *1*(2–4). doi:10.1038/s41893-017-0013-9

Ahuja, S., Nikolova, N., & Clegg, S. (2017). Paradoxical identity: The changing nature of architectural work and its relation to architects' identity. *Journal of Professions and Organization*, *4*(1), 2–19.

Åkerstrøm Andersen, N., & Stenner, P. (2019). Social immune mechanisms: Luhmann and potentialization technologies. *Theory, Culture & Society*, *37*(2), 79–103. doi:10.1177/0263276419868768

Archinect. (2004, Jul 27). *Peter Eisenman: "Liberal Views Have Never Built Anything of any Value."* Retrieved from https://archinect.com/features/article/4618/peter-eisenman-liberal-views-have-never-built-anything-of-any-value

Bogman PalmJaguar. (n.d). *Bogman Palmjaguar outlaw rock gardener of 1%.* Retrieved from http://www.palmjaguar2.com

Brenner, N. (2019). *New urban spaces: Urban theory and the scale question*. Oxford: Oxford University Press.

Brott, S. (2019). *Digital monuments: The dreams and abuses of iconic architecture.* London: Routledge.

Cairns, S. (2018). Debilitating city-centricity: Urbanisation and urban-rural hybridity in Southeast Asia. In R. Padawangi (Ed.), *Routledge handbook of urbanization in Southeast Asia* (pp. 115–129). London: Routledge.

Castricum, S. (2017). When program is the enemy of function. Gender-nonconforming experiences of architectural space. *Architecture and Culture*, *5*(3), 371–381.

Crawford, R. H., Stephan, A., & Prideaux, F. (2021, November). The EPiC database: Hybrid embodied environmental flow coefficients for construction materials. Resources, Conservation and Re*cycling*, 1061058. doi:10.1016/j.resconrec.2021.106058

Critchlow, K. (1976). R. Buckminster Fuller. *Architectural Design*, *46*/*5*, 296–297.

Deamer, P. (2020). *Architecture and labor*. London: Routledge.

Dorignon, L. (2015, September 21). And the urban exploded. *Society and Space*. Retrieved from https://societyandspace.org/2015/09/21/and-the-urban-exploded-by-louise-dorignon

Dowrick, S. B. (2008). *Rainer Maria Rilke: Bearing witness*. Doctoral Dissertation, University of Western, Sydney.

Dutto, A. A. (2018). RE: This thing called theory. *Ardeth*, *3*, 236–241. doi:10.17454/ARDETH03.15

Eisenman, P., & Iturbe, E. (2020). *Lateness*. Princeton, NJ: Princeton University Press.

Forensic Architecture. (n.d). About. Retrieved from https://forensic-architecture.org/about/agency

Grose, M. (2017). *Constructed ecologies: Critical reflections on ecology with design.* New York, NY: Taylor & Francis.

Hagen, A. L. (2017). Sketching with knives: Architects & the confidence theory of magic. *Anthropology Today*, *33*(2), 24–27.

Ing, W. (2022, January 6). Foster + partners refuses to back RIBA sustainability report. *The Architects' Journal*. Retrieved from https://www.architectsjournal.co.uk/news/foster-partners-refuses-to-back-riba-sustainability-report?tkn=1

Iturbe, E. (2019). Architecture and the death of carbon modernity. *Log*, 47, 10–23.

Jon, I. (2021). *Cities in the anthropocene: New ecology and urban politics*. London: Pluto Press.

King, M. (1993). The 'truth' about autopoiesis. *Journal of Law and Society*, *20*(2), 218–236.

Latour, B., & Yaneva, A. (2012). Give me a gun and I will make all buildings move: An ant's view of architecture. In A. L. Harrison (Ed.), *Architectural theories of the environment: Posthuman territory* (pp. 107–115). London: Routledge; Taylor & Francis Books Ltd.

Lee, P. M. (2020). *Think tank aesthetics: Midcentury modernism, the cold war, and the neoliberal present*. Cambridge, MA: MIT Press.

Lilienfeld, R. (1978). *The rise of system theory: An ideological analysis*. New York, NY: John Wiley & Sons.

Lorch, R. (2017). Climate policies for buildings. *Building Research & Information*, *45*(5), 475–477.

Luhmann, N. (1987). The representation of society within society. *Current Sociology*, *35*(2), 101–108.

Malinowski, B. (1922/2002). *Argonauts of the Western Pacific: An account of native enterprise and adventure in the archipelagoes of Melanesian New Guinea*. London: Routledge.

Malinowski, B. (2017). Perils, magic and myths. In M. W. Young (Ed.), *The ethnography of Malinowski: The Trobriand Islands 1915–1918*. London: Routledge.

Maturana, H. R., & Varela, F. J. (1980). Autopoiesis: The organization of the living. In H. R. Maturana & F. J. Varela (Eds.), *Autopoiesis and cognition: The realisation of the living* (pp. 59–138). Dordrecht: Reidel Publishing Company.

McHale, J. (Ed.). (1967). 2000 +. *Architectural Design*. 37.

Morris, I., & Morris, I. (2015). *Foragers, farmers, and fossil fuels*. Princeton, NJ: Princeton University Press.

Nightingale, A. J., Eriksen, S., Taylor, M., Forsyth, T., Pelling, M., Newsham, A., … Whitfield, S. (2019). Beyond technical fixes: Climate solutions and the great derangement. *Climate & Development*, *12*(4), 343–352.

Plumwood, V. (1993). *Feminism and the mastery of nature*. London: Routledge.

Plumwood, V. (1995). Human vulnerability and the experience of being prey. *Quadrant*, *39*(3), 29–34.

Plumwood, V. (2012). *The eye of the crocodile*. Canberra: ANU Press.

Raisbeck, P. (2019). *Architecture as a global system: Scavengers, Tribes, Warlords and Megafirms*. Bingley: Emerald Publishing Limited.

Raisbeck, P. (2020). Architectural research as an autopoietic system: Shifting the focus from actors and identities to systems. In 36th Annual Conference, ARCOM, September 2020, 7–8. Working Papers, Glasgow. Retrieved from http://hdl.handle.net/11343/252825

Schumacher, P. (2011). Parametricism and the autopoiesis of architecture. *Log*, *21*, 62–79.

Scott, F. D. (2016). *Outlaw territories: Environments of insecurity/architectures of counterinsurgency*. Cambridge, MA: MIT Press.

Spencer, D. (2016). *The architecture of neoliberalism: How contemporary architecture became an instrument of control and compliance*. London: Bloomsbury Publishing.

Spencer, D. (2021). Going to ground: Agency, design and the problem of Bruno Latour. In D. Spencer (Ed.), *Critique of architecture* (pp. 150–162). Basel: Birkhäuser.

Steele, W. (2020). *Planning wild cities: Human–nature relationships in the urban age*. London: Routledge.

Steele, W., Wiesel, I., & Maller, C. (2019). More-than-human cities: Where the wild things are. *Geoforum*, *106*(November), 411–415. Retrieved from https://www.sciencedirect.com/science/article/abs/pii/S0016718519301174

Tafuri, M. (1987). *The sphere and the labyrinth* (P. d'Acierno & R. Connolly, Trans.). Cambridge, MA: MIT Press.

Tayob, H. (2019). Architecture-by-migrants: The porous infrastructures of bellville. *Anthropology Southern Africa*, *42*(1), 46–58.

Territorial Agency. (n.d.a). Territorial agency [website]. Retrieved from https://www.territorialagency.com/

Territorial Agency. (n.d.b). Territorial agency—museum of oil [website]. Retrieved from https://www.territorialagency.com/museumofoil

Trundle, A. (2020). Resilient cities in a sea of islands: Informality and climate change in the South Pacific. *Cities*, *97*, 102496.

Urban Theory Lab. (2015, March 17). Urban theory lab, "operational landscapes" exhibition at Melbourne school of design [website]. Retrieved from http://urbantheorylab.net/news/operational-landscapes-exhibition-in-Melbourne

Urban Theory Lab. (2021). Data-spheres of planetary urbanization [video]. Vimeo. Retrieved from https://vimeo.com/555245282

Varela, F. J., Maturana, H. R., & Uribe, R. (1974). Autopoiesis and the organisation of living systems, its characteristics and a model. *Biosystems*, *5*, 187–196.

Chapter 8

Conclusion: Reworlding the Canon

Introduction

This book can be read as a pilgrimage through an architectural landscape of magical mists that seem at odds with climate science. On emerging from these mists, an actual distance may seem to have been traveled: the architectural cities, the climate imaginaries of architects, appear somehow real – but these dreams are no more real than the circuits of social media and popular science that feed them. The climate imaginaries of architects are produced through theories, polemics, built and unbuilt projects, and performances in public such as lectures and social media. And after everything is said and done, we might ask: what now is a city given the climate emergency? Within architecture these dreams have many different – but perhaps not so different – autopoietic languages, creeds, and cults. These mists are part of a magic that appears not to see the Earth and the ecological violence of modernity. This magic produces and reproduces the idea of a good and kind Anthropocene: a future Anthropocene that is benign because, thanks to architects, it is in climate equilibrium. In these concealing mists the architects are protected from world of dangerous carbon emissions. However, the different dream cities created in these mists, are curiously deaf to the Earth itself, and this anacusis prevents architects from healing the planet.

To conclude, rather than dwelling on the dreams of architects, and to be more realistic, we might ask: How do architects engage with climate science and policy interfaces? As document herein this engagement is at best fragmentary. This question suggests a further question, to what extent are architects engaged with environmental politics and governance at any scale? An overview of the climate activism of architects at different scales – city, region, nation can globe – appears to be both scant and fragmented (Table 6.1). Architects are not represented – or seem much interested in – the SDGs and the NUA. Architects appear reluctant to adopt or discuss more radical actions and political tactics; beyond the niceties and polite statements of their membership organisations.

The absence of architects in civil society groups addressing climate change may only partly indicate why architecture is missing from SDG urbanization debates and theory. Indeed, as suggested herein, architects are also not present in the global discourse of urbanization, emanating from urban planning. While the

Architects, Sustainability and the Climate Emergency, 215–233

doi:10.1108/978-1-80382-291-420221008

poetic dreams of architects may be sufficient entertainment for architects themselves, these poetic dreams – and corresponding theories, prototypes, and spatial practices – are far removed from global climate and development debates in the urban sciences. In stressing their own knowledge autonomy, architects themselves are too easily seen by other urban scholars as simply being a specialized design discipline. Yet this relationship between architecture and other urban disciplines also has a reciprocal ambivalence. The elision of architecture in urban studies suggests a field of knowledge based on city scales – rather than the architectural scale – with expert domains neatly separated as "specialized academic disciplines" (Acuto, Parnell, & Seto, 2018). Urban studies, unlike architecture with its surfeit of climate imaginaries, appears to be in desperate need of disciplinary imagination (Phelps, 2021). In any case, architectural design, its ways of knowing, seeing and spatial practices should not be occluded in urban studies on sustainability and climate.

In turn, this points to how architecture constructs itself as a science. Because of the emphasis on design, traditional research methods are not often taught in architectural schools. Much has been written about design research and its trajectories since the advent of the first research assessment exercises in universities earlier in the Millennium (Raisbeck, 2019, pp. 87–90). It is difficult to know what to say about the way architects draw on popular science and conceptualize notions of scientific methodologies in their own design research. The pursuit by architects of a good and future-perfect Anthropocene has created a feed of climate imaginaries, based on the tropes of what architects perceive science to be: the laboratory – as an apolitical crucible; rhetorical dialectics – as scientific logic; basic science – signified by the superficial observation of nature; the inductive translation of geometries from nature — as a cypher for scientific method; prototyping – read as applied science and not innovation; material properties – treated as fact despite the ambiguities of carbon accounting; and so-called evidence-based research – black-boxed inputs and outputs.

The above litany of sins may be why Flora Samuel has asked "why is the architectural profession so confused about the nature of its knowledge" (Samuel, 2018, p. 74). Research methodologies are puzzling to architects – despite their claims to methodological certainty. The answer to this is not to be found in the coproduction of knowledge between clients and architects or in Post Occupancy Evaluation Studies (Samuel, 2018, pp. 191–200). In contrast to simplistic solutions, there are other perspectives about architectural expertise and applied research that could be opened up. These other perspectives, briefly noted here, would involve thinking about the profession's norms, logics and practices of research differently – not just seeing research as being design-centric. A few starting points: to see research and expertise as emerging from tacit, relational practices and systems and not from individuals but as "systems of meaning" (King, 1993). Self-reflexivity might be a part of such a renewal: architects might consider how they employ research vocabularies through "generating and distributing, but maintaining control over, words that others use to guide their thinking and behaviour" (Loewenstein, 2014, p. 65). Chan (2016) employs Bergson (1913/2001) to suggest that expertise needs to be conceptualized as a "process of becoming" – as compared to a process reinforcing disciplinary autonomy (p. 1).

Other Kinds of Magic

Feral Atlas

The vignettes of the previous chapter indicate that the magic of architects – and their little kits of spells – necessitate new kinds of magic. The two dominant modes of urban analysis – even in the digital era for architects: the ideogram urbanism à la Le Corbusier and the diagram urbanism à la the Smithson's, no longer seem relevant in the context of the climate emergency. Le Corbusier's ideograms are based in the practices of purism and elementary forms: a tree becoming a facade, the part shape of a body becoming a slab block and Modular man the ultimate abstract diagram. These translations – from ideogram to form – suggest the minor scale has been accounted for; instead, the figures of these bodies have been atrophied and incorporated into an abstract machine. Alison and Peter Smithson sought to revise Corbusier's ideograms in their Golden Lane Overlay infrastructure diagrams of 1952 – presented at the 9th CIAM conference in Aix-en-Provence (van Den Huevel, 2000, p. 44). Each of their diagram's layers (organic lines, branched trees, and objects in a field) suggested a fluid urbanism, overlaid one onto another in plan – Bernard Tschumi also springs to mind in this context. By the 1970s the Smithsons, after the social failure of the Robin Hood Gardens project had – unlike other architects – abandoned the idea of an infrastructural layering of cities: "a turn away from the solution to be universally consumed toward solutions personally made or chosen" (van Den Huevel, 2000, p. 50).

To expand this a little further, these important analytical and diagrammatic traditions in twentieth-century architecture are also evident in the work of bio-digital architects: through the idea that granular forms derived from biology can – through the miracle of parametric – endlessly expand, morph, and change into global patterns – in the Smithson's word's "to be universally consumed" (p. 50). Rather than counting for urban detail and micro-scales, these modes of analysis and design methods – often conflated together – erase minutiae: that erase what, the Smithson's might call "solutions personally made or chosen" (p. 50). New instruments of analysis, new kinds of magic, are now needed and must by necessity recognize ecological lifeworlds and more. In architecture, the involuntary drift and impetus to casually upscale to the global – a tendency grounded in both colonialism and modernity – is a dangerous thing. Every recourse to, and conceptualization of the global, by architects – as seen in previous chapters – appears to validate and signify concern for the planet as whole, and yet paradoxically erasing all in its path to further the ecological destruction of the Earth.

The Feral Atlas is another kind of magic: A magic that does not erase granularity or lose it in a process of global extrapolation. This is a magic that indicates the possibilities of developing new forms of spatial analysis suitable to interrogating the circumstances and entanglements of the Anthropocene. The Feral

Atlas is based on field reports from both scientists', humanists', and artists' maps. The Atlas thus suggests a different approach to the macrodiagrams of architects:

> …the ecological worlds created when nonhuman entities become tangled up with human infrastructure projects. Seventy-nine field reports from scientists, humanists, and artists show you how to recognise "feral" ecologies, that is, ecologies that have been encouraged by human-built infrastructures, but which have developed and spread beyond human control.
>
> (Feral Atlas, 2021a)

The Atlas itself is not simply a catalogue or a map. Or even a book that documents the ruinous. Instead, and unlike the biodigital architects fixed on parametric relationships, the Atlas:

> …draws on the relational potential of the digital to offer new ways of analysing – and apprehending – the Anthropocene; while acknowledging danger, it demonstrates how in situ observation and transdisciplinary collaboration can cultivate vital forms of recognition and response to the urgent environmental challenges of our times.
>
> (Feral Atlas, 2021a)

The Atlas is predicated on understanding "nonhuman entanglements and intersection with the human" (Feral Atlas, 2021a). The Atlas presupposes that humans and nonhumans shape each other – repositioning human agency rather than reducing it. The Feral Atlas is thus a different type of magic, to that magic produced by many architects; an effort to map the Anthropocene not simply in terms of the human but also in terms of the nonhuman: "Other species, as well as nonliving things, make it possible to be human" (Feral Atlas, 2021b). While the Feral Atlas presupposes a "spatial analysis" of the Anthropocene, it also situates this analysis in histories that show the granular effects of the Anthropocene (Feral Atlas, 2021b). The Atlas is a magnifying glass that seeks to expand detail, of materials and living beings, of situated relationships, rather than push these further away.

Urban Poverty

Another kind of possible – but not the only – magic is the magic related to an understanding of the subjectivities of urban poverty. In this regard, Pedro Pacheco's interest in the philosophy of the Catholic educationalist Paulo Freire and his work the *Pedagogy of the Oppressed* (2007) points to a range of liberatory theories that architects are by and large oblivious to. It might not be expected that architects would have any commonality with the educational theories or even the liberation theories of Catholic theologians. Nonetheless, these theories and practices are arguably much less esoteric than the shamanism and sorcery of global architectural firms. As a movement that arose from the political debates of the 1960s and 1970s, liberation theology has customarily been associated with Latin American theological scholars. For architects liberation theology highlights the paucity of their current theories and may offer new avenues of design activism. This is because of its links to political activism, peace studies, and coproduction practices in the Global South – links that are also inextricably linked to sustainability transitions. As a basic set of concerns liberation theology. As Okanlawon (2021) observes:

> Liberation theology has two basic principles: first, it recognises a need for liberation from any kind of oppression- political, economic, social, sexual, racial and religious oppression; second, it asserts that theology must grow out of the basic Christian communities and should not be imposed from above.
>
> (p. 2)

The point here is, if put somewhat cursorily: the urban sociology of poverty might expand the orthodox theories and instruments of those architects focused on rapid urbanization. Arguably the morphological studies, and complexity theories, of those architects seeking to explain the dynamics of urbanization in the Global South have reached a limit. This limit is evident when it can be seen that descriptive, diagrammatic, and graphic data mapping can only go so far in describing rapid and incessant urbanization. Cybernetic-orientated theories cannot explain everything in the world. The second broader point is: while liberation theology might appear to be an obscure field to draw on; this point indicates the limiting range of sources drawn on in normative architectural discourse.

The work of Mitlin and Satterthwaite (2012; Satterthwaite & Mitlin, 2013), alongside the community practices of liberation theology, suggests how a different kind of magic might arise through a consideration of urban poverty. In light of all the hyperbole about global data and the climate imaginaries expounded by architects, their warning should be salutary:

> We believe that the scale and depth of urban poverty is ignored within most low income nations, many middle-income nations and globally. We believe that this reflects a considerable misrepresentation and underestimation of urban poverty, and occurs because of the very narrow ways in which poverty is usually conceived, defined and measured.
>
> (p. 1)

It is not simply enough for architects of the Global North to cite the aggregate statistics of urbanization in the Global South. In contrast and by way of example, Mitlin's work has been connected to the circumstances of water and sanitation in the Global South: "it will be increasingly difficult for cities to keep pace with the growing demand for water" (Beard & Mitlin, 2021, p. 1). Mitlin's group, the Global Development Institute at the University of Manchester, have published related work regarding sanitation services in the Global South (Beard, Satterthwaite, Mitlin, & Du, 2022). This work examines sanitation provision gaps and the aspiration of SDG target 6.2 that hopes to "achieve access to adequate and equitable sanitation and hygiene for all and end open defecation by 2030" (p. 1). This research seems to have a distant echo in architectural theory with the significant contributions of Frichot's notion of dirty theory and McGaw's concepts of dark matter, as discussed in Chapter 2. This comparison: between the scatological in architectural theory and the green city, built up as a mosaic in Chapter 5, the city of a tamed Anthropocene created through an image-politics of technical and material innovations; is a city that never mentions fecal matter – in other words, shit – or sanitation as a problematic of urbanism. The same can also be said of the floating and polluting monsters of Chapter 3. Nonetheless, embedded in the cities of the Global South is another kind of magic that architects should take heed of. As Banks, Lombard, and Mitlin (2020) note:

> We suggest that reconsidering informality as a site of critical analysis offers a new perspective that draws on and extends political economy approaches, and helps us to understand processes of stratification and disadvantage. We seek to highlight the significance of the informal-formal continuum at the same time as challenging this dichotomy, and to explore emerging theoretical and empirical developments, including changing attitudes to informality; the increasing salience of agency; and informality as a strategy both for elite and subaltern groups.
>
> (p. 223)

The Dream of Schizoanalytic Architecture

Can architecture and architects produce new spells and new magic? Guattari in *Schizoanalytic Cartographies* devotes a short chapter to architecture entitled "Architectural Enunciation" – positioned between sections on "Jean Genet" and the "Polish theatre of Theatre of Witkiewicz." Pivoting from a reading of the French architectural historian Pierre Boudon, Guattari sets out a kind of architecture of the architectural subject: an architecture of architecture. In the introduction to the section Guattari states:

> Enunciation rests on a constantly modified subjectification. That is to say, one cannot connect it and disconnect it as one would a computer. One never stops arranging it and decomposing it so as to recompose it on different bases. From this point of view, theatre work may constitute one of its most significant paradigms.
>
> (Guattari, 1989/2013, p. 241)

In this section on "Architectural Enunciation," a framework of how architects might abandon static singularizations is outlined. The concept of collective assemblages of enunciation – discussed in *A Thousand Plateaus* – suggests how to fashion an architecture of collective enunciation and continual resubjectification (Deleuze & Guattari, 1988; Guattari, 1989/2013). In setting out this, they emphasize that a fixed, typological, and functional description of space, what Guattari calls "taxonomic modes of spatialisation" must be abandoned. Within this new field, this renewed architecture of architecture, Guattari (1989/2019) identifies eight types of assemblages: geopolitical, urbanistic, economic, functional, technical, signifying, existential territorialization, and a scriptural enunciation (p. 236). These types correspond to the collective voices that are both "virtual" and also "diachronic discursive components" because of their latent possibilities in architecture (p. 237). In relation to functional enunciation, Guattari quote's Virilio (1991):

> If there are monuments today, they are certainly not of the visible order, despite the twists and turns of architectural excess. No longer part of perceptible appearances nor of the aesthetic of the apparition of volumes assembled under the sun, this monumental disproportion now resides within the obscure luminescence of terminals, consoles and other electronic night-stands.
>
> (Guattari, 1989/2013, p. 234)

The above assemblages are further articulated as follows: The technical enunciation, collectively linked to "little by little, the entirety of scientific and technical disciplines" (p. 235); existential territorialization or spatialities seen as Euclidian spaces, projective spaces, and labyrinthine topological spaces. They then suggest that the body in space should be enunciated in an organic, "labyrinthine (or rhizomatic)" fashion (p. 236). And finally, Guattari (1989/2013) articulates the scriptural enunciation that:

> ...articulates the ensemble of other enunciative components. Because of the diagrammatic distance that it introduces between expression and content, and through the coefficients of creativity that it generates, architectural design promotes new potentialities, new Constellations of Universes of reference.
>
> (p. 236)

These new potentialities in this architecture of architecture also start with the ethico-aesthetic aspects of the built project. Guattari then explains these aspects as "existential dimensions" – referring to Bahktin's (1990) tripartite division of enunciation (cognitive, ethical, aesthetic). In Guattari's hands, this trio becomes three ethico-aesthetic ordinates: the "cognitive," "axiological," and "aesthetic" (p. 237). The cognitive is described as being "energetico-spatial-temporal" arising from the geopolitical, urbanistic, economic, functional, and technical voices – or assemblages as outlined above (p. 237). The second is the axiological, which is, validating and ethical, but perhaps better put in Guattari's (1989/2013) words:

> ...the ensemble of systems of anthropocentric valorisation, as much of an ethical as of an economic and political order.
>
> (p. 237)

These aesthetic ordinates encompass among other things "thresholds of consummation," of completion, and this enables a built project to "start to emit sense and form on their own count" and through a process of "signifying enunciation and existential deterritorialisation;" hence the "built, the lived and the incorporeal find themselves are articulated with one another" – and thus Guattari outlines how he views architecture: in this universe, an entity that is polyphonic and should not be reduced to the anyone particular magical spell.

Assemblages of enunciation have been explored by Mazzei (2017) through an interpretation of Deleuze and Guattari's concept of Minor Literature and the context of qualitative research – indicating the need for similar studies in architecture – and how these assemblages might help to fashion what they designate as minor inquiry. Arguably the work of Frichot, but not Spencer, is an attempt to pursue the idea of a minor register in architectural theory. As Mazzei (2017) interprets it an assemblage of collective enunciations:

> ...do not emerge from nothing, instead they are spoken from our bodies, our histories, our communities, our materialities. We don't just speak of our own past, but of the past in an encounter in which our own and others' voice flow and are entangled with other materialities.
>
> (p. 682)

The import of this concept is that a singular authorial, or architectural voice, or singular concept, where the relationship between content and expression is stable is untenable if architects seek change; and while architecture is composed of machinic assemblages, as Guattari points out, it can also be thought of collective assemblages of enunciation; in this fashion shifting the focus away from a single voice (Why do we keep needing to say this in architecture?). Hence for Guattari:

> Reinventing architecture can no longer signify the relaunching of a style, a school, a theory with a hegemonic vocation, but the recomposition of architectural enunciation, and, in a sense, the trade of the architect, under today's conditions.
>
> (p. 232)

Surprisingly, previous Guattari studies and architecture, as discussed in Chapter 2, have tended to avoid an analysis of architecture as a kind of minor literature and assemblages of collective enunciation. For architects, the larger scale and other bigger interpretive lenses always seem to prevail. For example,

Genosko (2015) focuses on Guattari and larger-scale issues of urbanism – the megamachine. Characteristically, Brott read's Guattari's project in psychoanalytic terms through the lens of Lacan. Gorny pursues the murky concept of the assemblage, and despite an interest in architecture's transformative capacities, their emphasis on material arrangements elides assemblages of collective enunciation. On the other hand, Abrahams (2019) interrogates Deleuze and Guattari's notion of machinic assemblages, and their use in architectural theory, but then does not discuss assemblages of collective enunciation. We might easily conclude that the above approaches to Guattari and architecture are too often conceptualized at larger scales but even this assertion is contingent in regard to architectural theory; and clearly these difficult and contested Guattarian concepts, of abstract machines and assemblages of collective enunciation, require further explication (Berressemm, 2020, pp. 187–193).

Now with some distance traveled in this book, we might, if somewhat broadly, consider Guattari's projects of schizoanalysis and ecosophy in relation to architecture and the climate emergency – noting that: too deep, and we drown; and too shallow, we only wade. In light of the climate emergency, how might we approach or consider the schizoanalytic in architecture? Guattari (1989/2013) goes on to say that:

> The architectural form is not called on to function as a gestalt closed in on itself, but as a catalytic operator triggering chain reactions at the heart of modes of semiotization that make us escape from ourselves and open us up to original fields of possibility.
>
> (p. 238)

And they then suggest that the project of architecture is "the recomposition of existential Territories, in the context of our societies devastated by capitalistic Flows." Architects are then required to achieve an "effect of existential consummation and superabundance of being" to be "carried off" by a process of "historical enrichment and re-singularization of desire and of values" (Guattari, 1989/2013, p. 239). If the project of Schizoanalytic Architecture is a project to reinvent architecture, to continually reshape architecture – alongside and through the climate emergency – then it is clear that architects require new kinds of magic. In many ways, Glowczewski's (2020) Map of Yam Dreaming is more architectural and more indicative of sustainability than any of the green projects in Dezeen's sustainability long list (Fig. 2.2).

Reworlding

In her book *Decolonising Solidarity: Dilemmas and Directions for Supporters of Indigenous Struggles*, Clare Land (2015) suggests how nonindigenous people might develop new frameworks supporting Indigenous struggles. Land argues that research is deeply implicated with processes of colonization and the appropriation of Indigenous Knowledge. Indigenous Knowledge and Knowledge Systems are central to understanding the climate emergency and the voice of the Earth: how then, might architectural theory and practice be decolonised?

One way of opening up a space of decolonization is from the viewpoint of the architectural archive and also, by implication, to consider the Earth itself as an archive. The International Council on Archives (ICA) Tandanya-Adelaide Declaration (TAD) recognizes and values Indigenous cultural expressions in both public and private archives (International Council on Archives, 2015). As the ICA declares, First Nations "have sustained their own evolving social, cultural, political and spiritual identities" and these identities exist within dynamic "Indigenous Knowledge systems and ways of knowing." As the ICA declaims, "Indigenous Peoples – their philosophies and knowledge models – have prevailed despite enveloping colonial programs of assimilation and genocide" (International Council on Archives, 2015, p. 1). The Declaration thus recognizes the trauma to First Nations individuals inherent within settler archives. For example, within Australian archives, First Nations trauma may be engendered through narratives related to frontier violence and massacres; the triumph and depiction of colonial power and the psychic violation and reframing of Indigenous Knowledge. The evidence of genocide and the presence of trauma exists within all archives, pointing to the need for architects, archival institutions, and custodians of collections to construct new relationships with First Nations communities (Land, 2015, p. 29).

However, architecture is generally associated with histories concerning particular architects and their projects. European and Global North histories of architecture have focused on colonial contexts, the development of cities, international movements, and the aesthetics of styles. Despite calls for inclusiveness, there is nonetheless at the heart of many architectural histories, a focus on colonization and settler progress, its pulses and momentum: an unrelenting focus on capitalist flows. The colonization of the Earth, and economic progress, lurks in the plethora of origin tales stuck to modern architecture from the early twentieth century to the present. The colonization of the Earth is often disguised in these histories and encoded through a fascination for European modernism. Contestable, as these broad statements might appear, the colonization and destruction of the Earth's lifeworlds, through architecture, including its contrails of ecological destruction, is rarely questioned in architectural history and theory.

Land (2015) suggests, in the context of First Nations, that research is a problematic exercise – and it might be added, so too is design research. This outlook is because of the implication of research in the colonization of Indigenous Peoples and the appropriation of Indigenous Knowledges (p. 26). Land argues that, at the outset of the research, there is a need for self-reflexivity that recognizes white privilege and that:

> To seriously develop non-indigenous research agendas would require accountability constructs to be established so that

> non-indigenous researchers were located as challengeable by Indigenous researchers.
>
> (p. 29)

In Australia, central to the question of decolonization and white complicity in research – including this author's privileged complicity – is the need for a process that, in Land's (2015) words, "Springs from the recognition by non-indigenous people that Australia is Aboriginal Land" (p. 230). In order to counter the onslaught of the Anthropocene and the stilted efforts by architects to imagine a good Anthropocene, it can be asked: what might the architect have to gain in engaging with architectural expressions of Indigenous Knowledge? How might Indigenous authority shape climate research and even the architectural design process itself? Linda Tuhiwai Smith, (2021) argues that decolonizing methodologies have several different modalities: "indigenising, intervening, reading, reframing and restoring" (Smith, 2021, p. 142). The first two of these modalities, humbly approached and perhaps only partly achieved herein, are the focus of the narrative described in this book: Indigenizing is the project of centering and recentering: "of the centering of the landscapes, images, languages, themes, metaphors and stories in the Indigenous world and the disconnecting of many of the cultural ties between the settler society and its metropolitan homeland" (Land, 29). Another modality is the project of intervention, a "proactive" project: "designed around making structural and cultural changes" (Smith, 2021, pp. 146–147). In Land's words this is a project with "clear structural concerns" and "strategic questions" addressing the question of "what will engage and maintain the contribution to meaningful social change by members of dominating groups" (2015, p. 29).

In short: sustainability transitions will not be affected without Indigenous Knowledge. This idea is particularly acute in Australia as the continent burns, knowing that it also contains the narratives of the oldest living culture on the planet. For example, the scholars of the Nulungu Research Institute advance a postqualitative research agenda that "foregrounds the sentient nature of life as ancestral power" (Poelina et al., 2022, p. 1). Moreover, postqualitative research foregrounds "Country's animacy and presence, revealing the essence of time as non-linear, cyclical and perpetual" (p. 1). For these scholarly voices, this is achieved through "*katiitjin bidi* or learning journey" (p. 1). Moreover, it is contended that:

> Today, global environmental changes are central matters of concern, with demand for climate care, species respect, recognition that the cry of the Earth and the cry of the poor are the same, and acknowledgement that human and planetary wellbeing is directly connected.
>
> (Poelina et al., 2022, p. 2)

In a recent issue of *Current Opinion in Environmental Sustainability* with the theme of "Indigenous Conceptualisations of Sustainability," the editors call for sustainability to be defined not simply as an overarching universal definition; nor should it be linked, in any way to the idea that the land is an object of economic growth (Virtanen, Siragusa, & Guttorm, 2020). Their voices point to new burgeoning trajectories of sustainability that account for:

> Deep-rooted relations with place and with resources encompass multiple ways to interrelate and connect with all beings, such as plants and nonhuman animals. This understanding draws from the idea that humans also depend on spiritual and nonhuman actors' agencies.
>
> (p. 80)

But it is not only the Earth that points to different architectural subjectivities. In collaborative research Mitchell et al. (2020) relate how the Yolŋu song spirals of Guwak and the SkyCountry articulate ideas of Indigenous Geography and Cosmologies; affirming that Indigenous People "have ongoing connections with and beyond the sky" in order to "speak back to promoters of space colonisation who frame their projects as harmless" (p. 1). SkyCountry is in stark contrast to the call by some for a "science of night" in urban studies as a way to further understand urban sustainability (Acuto, 2019). But SkyCountry is different to an urban, city-scaled, and seemingly rational and scientific conception of the night – and SkyCountry is also a kind of city if we are to think about it precisely. Mitchell et al. (2020) expands and reworlds our view of outer space by asserting that it is incorrect to assume:

> ...that none of the Indigenous people or beings who also live on Earth have travelled to or inhabited this space. Second, they assume that space is dead or non-sentient in itself, and that it is incapable of fostering life. Third, they understand that space is cleanly separated from Earth, meaning that what happens in space has no effect on Earth, or vice versa. (p. 1)

Given the above, it is time to turn to more authentic conceptions of magic, the magic that shadows each step of architectural sustainability. Where Indigenous Knowledge has been gazed on sightlessly, relocated, reframed, and resituated in the canonical archive, or is absent from the design process itself, all this does not mean that this knowledge no longer exists. The above research indicates that to heal the Earth and for architects to listen to its voice, projects of structural intervention and recentring can be pursued through the process of reworlding. Inexorably, in Australia people of white privilege – like this author – are starting to hear Country. We have not been able to hear for too long. The concept of reworlding makes space for Indigenous narratives rather than making and remaking settler, colonial, and eurocentric architectural narratives of violence. Through reworlding, new modalities of reading, reframing, and restoring Indigenous Knowledge in the climate imaginaries of architects will emerge.

The Uluru Kata-Tjuta Cultural Centre at Uluru, designed in collaboration with the local A̠nangu cultural community, was completed in 1995 (Gregory Burgess Architects, n.d). The Uluru project, calls architects to see anew. In this example, intangible knowledge of Country is embodied in two different processes of collective enunciation. Firstly, the Uluru project indicates how the nonphysical characteristics, practices, and expressions of Indigenous Knowledge and skills may be embodied within an architectural project. This project's design enunciates nonphysical and spatial representations, oral traditions, and linguistic structures and language. Secondly, the participatory design of the Uluru project indicates that architectural spaces can also be shaped and enunciated through music and the performing arts; social practices, rituals, and festive events; foods and clothing; and traditional craftsmanship (Janke, 2019, p. 159). In this project, it can be seen how Indigenous Knowledge, the voice of the Earth itself, exists in the surrounding landscape: this project is not an iconic settler shed. At Uluru, the gathering of Indigenous Knowledge and the process of collaboration was documented in the architect's narrative and different media, including images of the A̠nangu People and Uluru. This material is now present in the Gregory Burgess Archive at Avington Victoria and at Melbourne University in Australia (An audit of this material in participation with A̠nangu would determine what might be seen or exhibited). Through the Uluru project, knowledge and practices concerning Indigenous Cosmologies are enunciated in this project. This embodiment points to how ecology, the human and nonhuman, and living Indigenous Mythologies can exist, and be gathered, into a project within both a project and its corresponding archive, pointing to how such things are spoken to and heard in the broader architectural canon.

A manuscript received from Gregory Burgess entitled *Between worlds, before building: Reflections on living encounters in the sacred Country of Uluru* goes some way to documenting the Indigenous knowledge in the Uluru project (Anangu Knowledge as cited in Burgess, n.d). The commission began in early 1990 after the Mutitjulu community and the National Parks and Wildlife Service, working jointly, advertised for consultants to design and construct a cultural centre at Uluru in the Uluru Kata-Tjuta World Heritage National Park. As part of the Uluru process, it was decided that "the design process should allow 'unhurried time' for us to live on-site for a month with Anangu, to listen and consult with the Elders on the cultural 'lie of the land' and to establish a shared clarity of purpose and depth of understanding" (Anangu Knowledge as cited in Burgess, n.d). From the outset, the architect determined that:

> ...there were complex issues of establishing an appropriate process of collaboration and understanding: language, land/Country, culture and functionality.......What was needed here was a resonant belonging, a responsive process and building that wove together living – spiritual and physical – connections with Uluru, the sacred Country and its people.
>
> (Anangu Knowledge as cited in Burgess, n.d)

During the time he stayed in the Mutitjulu Community, the architect worked closely with an elder lawman who became the guiding figure in the collaborative process and the sharing of Tjukurpa. As related to the architect:

> *Tjukurpa* is the foundation of *Anangu* life and society. It has many complex but complementary meanings and refers to the creation period when ancestral beings created the world as we now know it but also refers to the present and future.
>
> (Anangu Knowledge as cited in Burgess, n.d)

As the community was consulted with "different aspects of the same stories were told and these began to build up a picture to the point where the whole perimeter of 'the Rock' was alive with Tjurkurpa events" (Anangu Knowledge as cited in Burgess, n.d). As a part of the process, it was decided to commission a number of paintings from community artists. By listening to Anangu and engaging with Tjukurpa, paintings and stories shaped, in a direct manner, the evolution of the building concept. Several Anangu people were asked to paint the Tjukurpa of Uluru. Interestingly, the architect also intuitively employed a dowsing pendulum to position each of the building's spaces. He wrote that this process "was a step of trust in my own intuition which also seemed to invite the Tjukurpa spirit into the conversation" and that "All these interweaving aspects were synthesised in the design process" (Anangu Knowledge as cited in Burgess, n.d). They recalled that:

> The cycles of ritual that have kept Tjukurpa alive and sustained at Uluru for many thousands of years have changed the land with the directed human consciousness of spirit, creating a sacred realm inhabited by ancestral beings.
>
> (Anangu Knowledge as cited in Burgess, n.d)

As they were to state unequivocally:

> The knowledge is in the land. You just have to listen. It's always been there and will always be there.
>
> (Anangu Knowledge as cited in Burgess, n.d)

To expand the argument a little more, through knowledge of Country, the Earth, cosmologies, and the nonhuman, both architectural projects, and cities can be reworlded. The Uluru project points to the need for architects to reworld their spatial practices. Reworlding can be defined as listening, hearing, recognizing, and folding in the Earth itself into the architectural canon. Reworlding locates and repositions the canon into a new realm, a realm where eurocentric concepts of space and time have collapsed. Reworlding recognizes individual Indigenous communities and their relationship to their Country. A realm where even in highly urbanized cities, the voice of the Earth might yet be heard, expanding the "global urban imagination" beyond limited policy interfaces (Acuto, 2022).

Through reworlding, the settler Anthropocene might be continuously resubjectivized by Indigenous Knowledge. This Indigenous Knowledge is not generalized but is specific to individual Indigenous Communities. Through reworlding, the settler frontier, rather than being defined as an edge of economic progress, is reclaimed, so that past and present human and nonhuman exist within different Cosmologies. But reworlding the canon is not simply achieved through recognizing the nonhuman as equal agents to humans within a dynamic ecological network. There is more: at Uluru, reworlding is a concept that places the ancient living beings of Tjukurpa as equal agents within its realm: and why not? As indicated above, for the Anangu at Uluru, this Cosmology is called Tjupurka (Anangu Knowledge as cited in Burgess, n.d).

In architectural projects, spatiality is commonly represented and present, as well as in the material, plans, images, or file documents that describe the project. The voice of the Earth itself is also connected to this spatial dimension, and it is through this spatial dimension that further dimensions and layers of Country are made evident. These other country dimensions might include oral histories and language, cultural expressions such as ceremonies, songs, and paintings. As Philip Jones documents in his work on the Australia Hillier map, the land is a map; everything has a name (Jones, 2002). A project, a city, and its housing, its landscapes and ecological systems will also be part of this dimension – as is the case at Uluru. In Australia, a landscape of Country, is also a landscape of a city, including seemingly inanimate objects such as rocks.

In beginning to recognize these elements, we can continue to pursue, and also expand, Clare Land's methodologies of decolonizsation: reading, reframing, and restoring (Land, 2015). Settler progress, colonization, and iconography are entrenched and widely distributed in Australian architectural history, as it is in any eurocentric history or canon. For instance, the work of Glenn Murcutt, Australia's only Pritzker prize winner, points to how the trope of the settler shed are melded deep into Australian architectural history (Davidson, 2006). For example, Murcutts' project of "environmental design" elides Country, and this is on display in Cynthia Davidson's 2006 review of Murcutt's work without any awareness of this absence (Davidson, 2006). As the Indigenous poet Evelyn Araluen who was born and raised on Dharug Country and a descendant of the Bundjalung Nation, writes in her poem *The Trope Speaks* about the ever-present inscription of colonial symbolism in settler culture (Araluen, 2020, p. 32).

Araluen's tropes are a reminder of Walter Benjamin's *Angelus Novus* looking back at the wreckage of modernity. In the hands of single auteur's intent on authorial agency, design knowledge is formed in a way that anticipates the future progression of time. This forward-looking chronology gives life to tropes of functional use, construction innovations, and symbolic capital in preparation for recognition – if not sanctification – within the architectural canon. Time past is either seen as a signifier of regression or as an ahistorical source of a sublime timelessness; a timelessness devoid of oral and cultural histories. These forward- and backward-looking tropes are then rewritten and rewritten through a process of mimesis in historical practices. In contrast, the description of design participation at Uluru indicates a process that has abandoned the forward- and backward-looking chronologies of modernity and its erasures.

Reworlding points to the need for architects and urbanists to consider abandoning forward propelling sequences and chronologies. But nonindigenous historians, such as myself, must also abandon methodologies that separate the Indigenous from the colonial past and the segmented accounts of architectural history, of divided human and nonhuman lifeworlds. Indigenous Knowledge is not an exotic and complex history to be dealt with separately or added to augment projects. Indigenous Knowledge cannot be parsed by architects or interwoven – dare we say assimilated – with other themes to create sustainable solutions. Concepts of settlement, and their histories, did not abruptly finish with the onset of colonization or at the inception of twentieth-century modernism.

Prevailing architectural theories and methodologies promote an architectural canon that creates an ever-increasing distance to the trauma of the frontier. This timing leads to another historiographic issue. How might archival practices, and even architectural history itself, enable architectural historians and theorists to abandon settler notions of modernity? How might new historiographic and theoretical approaches recognize colonial genocide and ecocide in architectural history – where is this new reworlded history? As white architectural theorists, we have all been complicit in not seeing the knowledge and spirit of Country in our writing and rewriting of settler history. As Wendy Steele has noted in her reading of Deborah Bird Rose's work, "We glamorise the frontier of civilisation, fetishise the violence of colonisation," and we are "blind to the ongoing livingness of Country" (Steele, 2019). Modernist violence is deeply embedded in architectural history. Reworlding is a concept that points to the need for new subjectivities, resingularizations, and expressions of collective enunciation.

Conclusion

Architects, through their climate imaginaries and magic, have created their own enchanted forest. Filled with theories, narratives, and projects, these forests are places of privilege magic and wonder. These are places of enchantment that have been created through the climate imaginaries of colonization and capitalism. Existing as they do within – and in a way, separate from – the Anthropocene, these forests, with their various enchantments, have created folkloric tales that both entertain but also blur the distinctions between architecture and popular science. These forests are places of liminality filled with: floating cities, algae or nature-inspired materials and supply chains, or construction prototypes or material innovations. For architects, these forests are thresholds between our current times and a living Anthropocene of the future. In these forests, architects hope that their various enchantments can become real. Yet, paradoxically, these enchanted forests are the products of a rapacious modernity that is destroying the lifeworlds of the Earth. In contrast, paradoxically, the lifeworld's of First Nations people – with their living mythological beings and ancestors – are actually more real. These lifeworlds indicate how cities should be questioned, reimagined and reworlded to counter dangerous anthropogenic warming.

Any discussion of sustainability and its political ecology in architecture in the present moment requires both critique and imagination. The story outlined in this book, beginning with Bogman PalmJaguar, Val Plumwood, Gosh's *Great Derangement*, and Felix Guattari should not be seen as being somehow curious or novel. Our imaginations need more than the strictures of academicism or neoliberalism. Architects can help in this regard – as seen here, the architectural spatial imagination can be compelling – but only if architects look to reworlding and speak for, and through collective voices. The planet is warming up alarmingly, and 2030 the time to limit this warming to 1.5°C is rapidly approaching: knowledge of history and poetry, a recognition of living mythological beings, no matter how tenuous these connections might seem, all suggest ways to not simply await climate catastrophe, a waiting that is limited by the moribund semi-theorizations of architects. The resultant political ecology constructed here allows us to reflect on how architects have engaged with, and might be engaged against the continuing degradation of the Earth. Many architects have good intentions about shifting the current trajectory of Anthropogenic warming. Many now question if their efforts to avert a climate emergency have fallen into a basket of good intentions with little effect. Perhaps, within architecture, there has been too much well-meaning effort for too little gain. In the future, climate activism will need to be more than the good intentions and green theatrics. The theoretical and sociotechnical instruments that architects have employed thus far to reduce carbon emissions are exhausted. This exhaustion is why all that has been done to date by architects is the creation of magical mists and enchanted forests.

References

Abrahams, G. (2019). The building as a Deleuzoguattarian Strata/Machinic assemblage. *Architectural Theory Review*, *23*(3), 363–379.

Acuto, M. (2019, December 18). We need a science of the night. *Nature*, *576*, 339. Retrieved from https://www.nature.com/articles/d41586-019-03836-2

Acuto, M. (2022). *How to build a global city: Recognising the symbolic power of a global urban imagination*. Ithaca, New York: Cornell University Press.

Acuto, M., Parnell, S., & Seto, K. C. (2018). Building a global urban science. *Nature Sustainability*, *1*(1), 2–4. Retrieved from https://www.nature.com/articles/s41893-017-0013-9

Anangu Knowledge. In Burgess, G. non-indigenous. (n.d). *Between worlds, before building: Reflections on living encounters in the sacred country of Uluru*. [Unpublished manuscript].

Araluen, E. (2020). *Dropbear*. Brisbane, QLD: University of Queensland Press.

Bakhtin, M. (1990). *'The problem of content, material, and form in verbal art' in art and answerability*. University of Texas Press.

Banks, N., Lombard, M., & Mitlin, D. (2020). Urban informality as a site of critical analysis. *Journal of Development Studies*, *56*(2), 223–238.

Beard, V. A., & Mitlin, D. (2021). Water access in global South cities: The challenges of intermittency and affordability. *World Development, 147*, 105625. doi:10.1016/j.worlddev.2021.105625

Beard, V. A., Satterthwaite, D., Mitlin, D., & Du, J. (2022). Out of sight, out of mind: Understanding the sanitation crisis in global South cities. *Journal of Environmental Management, 306*, 114285. doi:10.1016/j.jenvman.2021.114285

Bergson, H. (1913/2001). *Time and free will: An essay on the immediate data of consciousness* [F.L. Pogson, Trans.] (3rd ed.). London: George Allen and Company.

Chan, P. W. (2016). Expert knowledge in the making: Using a processual lens to examine expertise in construction. *Construction Management & Economics, 34*(7–8), 471–483.

Davidson, C. (2006). Raised to observe: Glenn Murcutt. Log No. 8: Toward a critique of sustainable architecture and landscape. Summer, pp. 31–39.

Deleuze, G., & Guattari, F. (1988). *A thousand plateaus: Capitalism and schizophrenia* (B. Massumi, Trans.). London: Athlone Press.

van Den Huevel, D. (2000). The diagrams of team 10. *Daidolos: Diagrammania, 74*, 64–67.

Feral Atlas. (2021a). *Feral Atlas: The more-than-human anthropocene*. Stanford University Press. Retrieved from https://feralatlas.org/

Feral Atlas. (2021b). *Introduction to feral Atlas*. Stanford University Press. Retrieved from https://feralatlas.supdigital.org/?cd=true&bdtext=introduction-to-feral-atlas

Freire, P. (2007). *Pedagogy of the oppressed*. New York, NY: Continuum.

Genosko, G. (2015). Megamachines: From Mumford to Guattari. Explorations in media. *Ecology, 14*(1–2), 7–20.

Glowczewski, B. (2020). *Indigenising anthropology with Guattari and Deleuze*. Edinburgh: Edinburgh University Press.

Gregory Burgess Architects. (n. d). Uluru Kata-Tjuta cultural centre. Retrieved from http://www.gbarch.com.au/projects/1995/uluru-kata-tjuta-cultural-centre/

Guattari, F. (1989/2013). *Schizoanalytic cartographies*. New York, NY: Bloomsbury.

International Council on Archives. (2019, October 5). *Tandanya – Adelaide declaration [report]* (p. 1). Retrieved from https://www.ica.org/sites/default/files/tandaya_adelaide_declaration_eng.pdf

Janke, T. (2019). *True tracks: Indigenous cultural and intellectual property principles for putting self-determination into practice*. Australian National University. Doctoral Dissertation. Retrieved from http://hdl.handle.net/1885/156420

Jones, P. (2002). Naming the dead heart: Hillier's map and Reuther's Gazetteer of 2 468 placenames in Northeastern South Australia. In L. Hercus, F. Hodges, & J. Simpson (Eds.), *The land is a map: Placenames of indigenous origin in Australia* (pp. 187–200).

King, M. (1993). The 'truth' about autopoiesis. *Journal of Law and Society, 20*(2), 218–236.

Land, C. (2015). *Decolonising solidarity: Dilemmas and directions for supporters of indigenous struggles*. Bloomsbury. London: Zed Books.

Loewenstein, J. (2014). Take my word for it: How professional vocabularies foster organising. *Journal of Professions and Organization, 1*(1), 65–83. doi:10.1093/jpo/jot004

Mazzei, L. A. (2017). Following the contour of concepts toward a minor inquiry. *Qualitative Inquiry, 23*(9), 675–685.

Mitchell, A., Wright, S., Suchet-Pearson, S., Lloyd, K., Burarrwanga, L., Ganambarr, R., … Maymuru, R. (2020). Dukarr lakarama: Listening to Guwak, talking back to space colonization. *Political Geography*, *81*, 102218. doi:10.1016/j.polgeo.2020.102218

Mitlin, D., & Satterthwaite, D. (2012). *Urban poverty in the Global South: Scale and nature*. London and New York: Routledge.

Okanlawon, S. O. (2021). Liberation theologies of the twentieth century: Insights for integral development of africa in the twenty-first century, *7*(1), 1–18. doi:10.38159/erats.2021711

Phelps, N. A. (2021). *The urban planning imagination: A critical international introduction*. New York, NY: John Wiley & Sons.

Poelina, A., Wooltorton, S., Blaise, M., Aniere, C. L., Horwitz, P., White, P. J., & Muecke, S. (2022). Regeneration time: Ancient wisdom for planetary wellbeing. *Australian Journal of Environmental Education*, 1–18.

Raisbeck, P. (2019). *Architecture as a global system: Scavengers, Tribes, Warlords and Megafirms*. Bingley: Emerald Publishing Limited.

Samuel, F. (2018). *Why architects matter: Evidencing and communicating the value of architects*. London and New York: Routledge.

Satterthwaite, D., & Mitlin, D. (2013). *Reducing urban poverty in the Global South*. London and New York: Routledge.

Smith, L. T. (2021). *Decolonising methodologies: Research and indigenous peoples*. London: Bloomsbury.

Steele, W. (2019). *Planning wild cities: Human–Nature relationships in the urban age*. London: Routledge.

Virilio, P. (1991). *Lost dimension*. Los Angeles, CA: Semiotext[e].

Virtanen, P. K., Siragusa, L., & Guttorm, H. (2020). Introduction: Toward more inclusive definitions of sustainability. *Current Opinion in Environmental Sustainability*, *43*(April), 77–82. doi:10.1016/j.cosust.2020.04.003

Index

Printed and bound by CPI Group (UK) Ltd, Croydon, CR0 4YY

30/10/2025

14763577-0001